ENVIRONMENT AND MARGINALITY IN GEOGRAPHICAL SPACE

Environment and Marginality in Geographical Space
Issues of land use, territorial marginalization and development in the new millennium

Edited by

ROSER MAJORAL
University of Barcelona, Spain

HEIKKI JUSSILA
University of Oulu, Finland

FERNANDA DELGADO-CRAVIDÃO
University of Coimbra, Portugal

LONDON AND NEW YORK

First published 2000 by Ashgate Publishing

Reissued 2018 by Routledge
2 Park Square, Milton Park, Abingdon, Oxon OX14 4RN
711 Third Avenue, New York, NY 10017, USA

Routledge is an imprint of the Taylor & Francis Group, an informa business

Copyright © Roser Majoral, Heikki Jussila, Fernanda Delgado-Cravidão and Individual Authors 2000

All rights reserved. No part of this book may be reprinted or reproduced or utilised in any form or by any electronic, mechanical, or other means, now known or hereafter invented, including photocopying and recording, or in any information storage or retrieval system, without permission in writing from the publishers.

Notice:
Product or corporate names may be trademarks or registered trademarks, and are used only for identification and explanation without intent to infringe.

Publisher's Note
The publisher has gone to great lengths to ensure the quality of this reprint but points out that some imperfections in the original copies may be apparent.

Disclaimer
The publisher has made every effort to trace copyright holders and welcomes correspondence from those they have been unable to contact.

ISBN 13: 978-1-138-73940-6 (hbk)
ISBN 13: 978-1-138-73938-3 (pbk)
ISBN 13: 978-1-315-18414-2 (ebk)

Contents

List of figures	ix
List of tables	xiii
Contributors	xvii
Preface	xix

1 Introduction   1
  Heikki Jussila and Roser Majoral

PART 1 — LAND USE AND ENVIRONMENT

2 Land use and abuse: On the ecological and spiritual   7
  marginalization of land
  Walter Leimgruber

3 Trajectories of change in marginal and critical regions in   25
  Southeast Asia and Southern China
  R.D. Hill

4 Economic scenario of the reclamation and utilisation of marginal   35
  lands in India
  Abha Lakshmi Singh

5 The problem of decision areas within a fragile ecosystem: Uco   51
  Valley, Mendoza, Argentina
  Gladys Molina de Buono

6 Agro-industrial enterprises and environmental fragility   61
  *Maria Josefina Gutiérrez de Manchón and Maria Estela
  Furlani de Civit*

7	A place in globalization: Land use in the Argentinean Pampa *Stella Maris Shmite de Castell*	72
8	The globalization impact on a marginal agricultural area *Gladys Mabel Tourn*	80
9	Transforming the fringe: Tobacco-related wood usage and its environmental implications *Helmut Geist*	87

PART 2 — TERRITORIAL MARGINALIZATION

10	The concept of territorial marginality: A reflection from the standpoint of the geographic image of Portugal at the end of the 20th Century *João Lúis Fernandes*	121
11	Territorial positionament of a peripheral region in the Atlantic Arch, Galicia (Spain) *María José Piñeira Mantiñán and Román Rodriguez González*	133
12	Reality and perception of marginality: The case of rural Catalonia *Márti Cors*	144
13	Territorial marginalization: Regional borders and globalization, some examples from Spain *Hugo Capella and Jaume Font-Garolera*	155
14	Political boundaries and the development of marginal areas: The border between Catalonia and Aragón (Spain) *Joan Tort í Donada*	169
15	Microperiphery in Mega-Cities *Thomas Blom*	178
16	Space, behaviour and marginality *Paulo Nossa*	194

PART 3 — DEVELOPMENT IN MARGINS AND PERIPHERIES

17 A pink invasion into the Dutch periphery 201
    Jan H.M. Maas and Johan Wisserhof

18 The changing agrarian structure of Israel's southern periphery 219
    David Grossman and Hanna Moshayov

19 Coping with business crisis in the Russian North: Strategies against turmoil 232
    Markku Tykkyläinen

20 Transition in the post-socialist countryside: The restructuring of rural settlements in Hungary and Russian Karelia 247
    Eira Varis

21 Residential and settlement issues of forest sector communities in the Republic of Karelia, Russia 263
    Minna Piipponen

PART 4 — SUMMARY AND CONCLUSIONS

22 Summary and conclusions 283
    Roser Majoral and Heikki Jussila

List of figures

Figure 2.1	A systemic approach to land use	9
Figure 2.2	The value system	10
Figure 2.3	World views since the Middle Ages	12
Figure 2.4	Traditional way of viewing marginality	19
Figure 2.5	A new way of viewing marginality	19
Figure 5.1	Territorial fragmentation: multiple influences at local level	54
Figure 5.2	Control and government institutions, Uco Valley, Mendoza, Argentina	58
Figure 6.1	The study region: topographical profiles	62
Figure 6.2	The study area: Uco Valley, Mendoza, Argentina	66
Figure 6.3	A systemic approach toward vine and wine industry	70
Figure 10.1	Globalization and territorial marginality	123
Figure 10.2	Marginal territories in Portugal, according to the enquiry	126
Figure 10.3	Municipalities with a better quality of life, according to the enquiry	128
Figure 10.4	Municipalities preferred for future residence, according to the enquiry	129
Figure 10.5	The best municipalities to spend a weekend break, according to the enquiry	130

Figure 11.1	Galicia's location in Europe and in the Iberian Peninsula, provincial and municipal division of Galicia	134
Figure 11.2	Portuguese bank branches in Galicia	137
Figure 11.3	Map of population density in 1996	138
Figure 13.1	Location map	156
Figure 13.2	The enclave of Llívia	159
Figure 13.3	The enclave of the Condado de Treviño	161
Figure 13.4	The enclave of Ceuta	163
Figure 13.5	The enclave of Melilla	164
Figure 13.6	The Rock of Gibraltar	166
Figure 15.1	The number of inhabitants in New York City's five boroughs	184
Figure 15.2	Changes in the population of New York City's five boroughs 1970-1990	185
Figure 15.3	A comparison of Manhattan's twelve districts on the basis of three variables	188
Figure 15.4	A comparison between the proportion of white population and the proportion receiving income support	189
Figure 16.1	Adults and children living with HIV/AIDS, end of 1997	197
Figure 17.1	The study regions in the Netherlands	202
Figure 17.2	Millions of pigs in the Netherlands, 1900-1998	204
Figure 18.1	Land use Eshkol Regional Council, 1998 season	223

Figure 18.2 Main citrus growing zones in Israel, 1992 and 1996, in dunams — 224

Figure 19.1 The interplay of human agency and structures — 236

Figure 19.2 Downward shift in costs and increasing demand in growth economies and upward shift in costs due to declining (mass) production and decreasing demand due to decreasing incomes in declining economies — 240

Figure 19.3 The closure and restart of economic and social activities — 241

Figure 19.4 Revenues and components of cost in old and new companies — 242

Figure 20.1 Factors of transition in a locality — 249

List of tables

Table 2.1	Organic farming in Switzerland in the 1990s	21
Table 4.1	Cost of reclaiming one acre of saline soils in Aligarh District, India	40
Table 4.2	Cost of reclaiming one acre of saline-alkali soils in Aligarh District, India	41
Table 4.3	Cost of reclaiming one acre of alkali soils in Aligarh District, India	42
Table 4.4	Utilisation and benefits of reclamation in Gursikran and Mahuakhera villages of Aligarh District	44
Table 5.1	Institutions and territorial influence of the various departments of the Uco Valley	56
Table 6.1	Vineyards, in ha, 1971, 1988 and 1996	63
Table 6.2	Fruit tree area, in ha, 1988	63
Table 6.3	Fruit tree area, in ha, 1996	63
Table 7.1	Gross regional agrarian product by branches, (%)	73
Table 7.2	Cereals and oilseed production - La Pampa - 1995	75
Table 7.3	Evolution of the production volumes of main crops (in tons)	76
Table 8.1	Province of La Pampa, comparison in 1895 and 1920	81
Table 8.2	Agricultural production and yield (kg/ha)	83

Table 9.1	World tobacco production, 1961 to 1998 (3-years' averages, for 1997/98 2-year average)	89
Table 9.2	Curing fuel as a percentage of flue tobacco cured, 1993	93
Table 9.3	Dimensions of wood-based curing of flue, 1991/95	94
Table 9.4	Individual country sources of energy for curing flue	95
Table 9.5	Percentage of flue growers who have woodfuel plantings	98
Table 9.6	Rates of fuelwood usage by fire-cured tobacco	102
Table 9.7	Rates of fuelwood usage by flue-cured tobacco in Africa	103
Table 9.8	Rates of fuelwood usage by flue-cured tobacco in Asia	104
Table 9.9	Rates of fuelwood usage by flue-cured tobacco in Latin America	105
Table 9.10	Rates of polewood usage in naturally cured tobaccos	105
Table 9.11	Annual wood consumption in tobacco farming, 1991/95 (stacked cubic metres per ton of cured tobacco)	107
Table 9.12	Fuelwood uses in rural industries of the developing world	108
Table 9.13	Global annual tobacco production using wood in 1991/95 (in '000 metric tons of dry weight)	109
Table 9.14	Global annual wood requirements of tobacco in 1991/95 (in '000 metric tons of solid wood)	110
Table 10.1	Terms associated with the concept of marginality	127
Table 11.1	Intervention axis of the Structural Funds for the period 1994-1999, in millions of pesetas; 166.386 pesetas = 1 €	140

Table 12.1	Real and perceived characteristics of the *comarca* of Solsonès	152
Table 17.1	Development of pig farming 1980-1997, (1980=100)	212
Table 20.1	Factors of transition in the case study villages at the general and sectoral levels	250
Table 20.2	Factors of transition in the case study villages at the local and individual levels	253
Table 20.3	Old and new actors and survival strategies in Hunya and Koivuselkä	257
Table 21.1	Population of Pulchejla village soviet by nationality, (%)	269
Table 21.2	Satisfaction with living conditions in Koivuselkä	270
Table 21.3	Interviewees' opinions about the most important problems in Matrosy	270
Table 21.4	Ownership of residence in Koivuselkä, (%)	272
Table 21.5	Ownership of residence in Matrosy, (%)	273
Table 21.6	Residents' opinions about the future of the home village, (%)	275

Contributors

THOMAS BLOM	University of Karlstad, Karlstad, Sweden
HUGO CAPELLA	University of Barcelona, Barcelona, Spain
MÁRTI CORS	University of Barcelona, Barcelona, Spain
FERNANDA DELGADO-CRAVIDÃO	University of Coimbra, Coimbra, Portugal
JOÃO LÚIS FERNANDES	University of Coimbra, Coimbra, Portugal
JAUME FONT-GAROLERA	University of Barcelona, Barcelona, Spain
MARIA ESTELA FURLANI DE CIVIT	CONICET, CRICYT, Mendoza, Argentina
HELMUT GEIST	LUCC-IGBP/IHDP, UCL-GEOG, Louvain La Neuve, Belgium
R.D. HILL	University of Hong Kong, Hong Kong, China
DAVID GROSSMAN	Bar-Ilan University, Israel
MARIA JOSEFINA GUTIÉRREZ DE MANCHÓN	CONICET, CRICYT, Mendoza, Argentina
HEIKKI JUSSILA	IGU, Comm.Dyn.Marg.Critical.Reg., Pisa, Italy
WALTER LEIMGRUBER	University of Fribourg, Fribourg, Switzerland
JAN H.M. MAAS	University of Nijmegen, Nijmegen, Netherlands
ROSER MAJORAL	University of Barcelona, Barcelona, Spain
GLADYS MOLINA DE BUONO	CONICET, CRICYT, Mendoza, Argentina

HANNA MOSHAYOV	Bar-Ilan University, Israel
PAULO NOSSA	University of Minho, Guimaraes, Portugal
MINNA PIIPPONEN	University of Joensuu, Joensuu, Finland
MARÍA JOSÉ PIÑEIRA MANTIÑÁN	University of Santiago, Santiago de Compostela, Spain
ROMÁN RODRIGUEZ GONZÁLEZ	University of Santiago, Santiago de Compostela, Spain
STELLA MARIS SHMITE DE CASTELL	National University of La Pampa, Santa Rosa, La Pampa, Argentina
ABHA LAKSHMI SINGH	Aligarh Muslim University, Aligarh, India
MARKKU TYKKYLÄINEN	University of Joensuu, Joensuu, Finland
GLADYS MABEL TOURN	National University of La Pampa, Santa Rosa, La Pampa, Argentina
JOAN TORT Í DONADA	University of Barcelona, Barcelona, Spain
EIRA VARIS	University of Joensuu, Joensuu, Finland
JOHAN WISSERHOF	University of Nijmegen, Nijmegen, Netherlands

Preface

In connection with the IGU Regional Conference, in Lisbon August 1998, the Commission on Dynamics of Marginal and Critical Regions met at the University of Coimbra, August 24-29. Following the Commission four year programme established in The Hague during the International Congress (1996) the theme announced for this meeting was 'Consequences of Globalization and Deregulation on Marginal and Critical Region Economic Systems' and of the large number of contributions 20 are presented in this 'Environment and Marginality in Geographical Space' book of the 'Spatial Aspects of Marginality'-series being published by Ashgate in association with the IGU Commission on Dynamics of Marginal and Critical Regions. The two previous books of this series are: 'Perceptions of Marginality' (1998) and 'Marginality in Space – Past, Present and Future' (1999).

This book on 'Environment and Marginality in Geographical Space' discusses the issue of marginalization from the point of view of the environment by using the viewpoints of land-use, landscape and development. The aim of this book is to give the reader an overview of the issues at hand, while keeping a close connection with practical real world examples of what, where and how environmental issues manifest on marginal areas.

The editors would like to thank all the contributing authors for their work, and thanks are also due to the organisers of the Coimbra meeting that made this book possible. Although, there are many other people we would like to mention, special thanks are due to professor Maria Andreoli (University of Pisa) and to Hugo Cappella (University of Barcelona) and to João Fernandes (University of Coimbra) who have helped us in putting things in order and also pointed out mistakes that otherwise could still have existed in this final version.

Finally, we would like to thank Ashgate for the book series that has enabled the commission to publish much more of its work than initially was thought.

Barcelona and Pisa, September

Roser Majoral and Heikki Jussila

1 Introduction

HEIKKI JUSSILA AND ROSER MAJORAL

The question of the quality of the environment has grown in importance during the 1990s. The rightful concerns about the issues of global warming and global change have increased the importance of research dedicated to these issues. This book on the issue of *Environment and Marginality in Geographical Space* looks at the issues of environment and land-use as well as those of territorial marginalization and development from environmental point of view. The attempt of this book is to analyse the many faceted image of environment in geographical space.

This book is the fourth in the *Spatial Aspects of Marginality*-series of books produced from the research work done under the programme of the IGU Commission on Dynamics of Marginal and Critical Regions. The aim of this book is to develop and give new visions for the readers about the marginal regions of the world. This book is the outcome of a scientific meeting of the IGU Commission on *Dynamics of Marginal and Critical Regions* held in Coimbra in August 1998 in occasion of the IGU Regional Conference held in Lisbon. The conference had two themes that focused on implications of 'Globalization in Geographical Space' and on 'Environmental Issues in Geographical Space'. This is the second book being published from the conference, the first being that on the issues of Globalization in Geographical Space.

This book is divided into fours main parts:

a) Land use and environment,
b) Territorial marginalization,
c) Development in margins and peripheries and
d) Summary and conclusions.

The first main part of this book *Land Use and Environment* contains eight chapters that analysed the different aspects of land use and environment. The issues discussed range from those of 'philosophical and ethical aspects' put forward in the first chapter by Leimgruber (Land use and abuse: On the ecological and spiritual marginalization of land) to the issues regarding globalization and the effects it has had on the land use in a specific area, the case of Argentinean Pampa, as discussed by Molina de Buono, Gutiérrez de Manchón and Furlani de Civit, Shmite de Castell and Tourn. The other issues put forward in the first part discuss aspects of time in land-use; the chapter by Hill on trajectories, the chapter by Singh on the over use of land, in the case of India and finally the chapter by Geist on the use of wood in case of tobacco growing countries.

The second main part of this book, *Territorial Marginalization*, contains seven chapters that analyse changes in the aspects of territorial marginalization. The first article in this part by Fernandes looks at the concept of territorial marginality and he emphasises how difficult it is to concretise the term and the criteria to be used to define the concept. The article of Fernandes is followed by the articles of Piñeira and Rodríguez and by Cors that deal with the territorial marginality in Spain.

The second part then continues to address the issue of borders as marginal territories in Spain. The first of these chapters by Capella and Font look at the impact of border to a region. They emphasise the importance to different types of elements that can reinforce the process of marginalization or reverse this process. The role of environment as a physical obstacle is looked into as well as the fact that through new political situations the physical 'limitations' can be overcome. The second article by Tort looks at a special case within Spain that has through political decisions created marginality within an area.

The chapter of Blom on Microphery in Megacities turns the issue of territory into the micro level through the analysis of a large city region. The question of territory in this context is smaller and closer to an individual. The question of environment of the problem in hand is that inner city problems seem to exist irrespective of the size of the city region in concern.

The last chapter of this part is by Nossa. His approach toward territorial marginalization is 'global' in the sense that he looks at the issue of AIDS/HIV at the level of continents. In this respect the territorial marginalization is apparent, and the most 'marginalized' region is the continent of Africa, where most of the AIDS victims can be found. The territorial aspect is also linked into an environmental aspect, which in this case is that of poor economic and social conditions. His approach enlarges the scope of the concept of environment and territory discussed in the first chapters of the two main parts of this book.

The last main part of this book discusses the issues of environment and territory from the point of view of 'development' under the heading of *Development in Margins and Peripheries*. This section contains five chapters that discuss and analyse the aspects and effects of development in peripheries and margins. The scope of the chapters in this section range from the issues of alternatives in regional development by Maas and Wisserhof and by Grossman and Moshayov and by Tykkyläinen. The last two chapters of this section and the book by Piipponen and by Varis discuss issues of local development and settlement. The discussion in this section concentrates on economic issues.

The articles in this book on *Environment and Marginality in Geographical Space* analyse the issue of environment from different angles and points of view. Despite of the 'heterogeneity' of the approaches presented they do, however, have common points of view. The authors of the articles in this book all have a great concern regarding the possibilities of marginal regions to cope in the current globalizing world. The articles in this book give insights to the reader how geographers from different parts of the world see the questions of environment, culture and economic development that all at the end are intertwined into a whole, the geographical space we live in. The articles do not have patented solutions to the problems they present, but – all in all – they do show ways to solve some of them. It remains the task of the reader to determine if these solutions present suitable models for solving those questions that the reader of this book might have.

Part 1 – Land use and environment

2 Land use and abuse:
On the ecological and spiritual marginalization of land

WALTER LEIMGRUBER

Introduction

Like any other system, land use systems are characterised by input, throughput, output, and feedback. Land, 'the basic natural resource' (Mather, 1986, p.1), is the stage of human activities, where raw materials are mined and transformed, and where waste is dumped. Mining and dumping are the two key operations at either end of the production chain, and both occur on legal and on unauthorised sites. Both are also a mirror of man's attitude towards land.

This chapter is about the relationship between humans and the land; its aim is to call for a new attitude towards land. Such a shift in perception is imperative, for since the industrial revolution, land has become marginalized in our minds. By regarding it simply as a resource for increasing economic profit, we have forgotten that land also needs care. Monocultures in farming, industrial activities using dangerous materials and military actions (Simmons, 1989, p.314) all result in the degradation and even destruction of land. Nuclear power stations which have been closed down, illegal waste dumps, craters left by artillery shells in Northern France and still visible eighty years after the end of World War I, nuclear test areas around the world, the defoliated Vietnamese forests, old Soviet military camps in former satellite states (Jauhiainen and Lootus, 1998) – they all bear witness to the destructive force inherent in mankind. The ecosystem may repair some of the damage, but the time frame will exceed human life expectancy. The cost to clear such destruction is prohibitive, as Jauhiainen, (1997, p.124) points out with reference to a former Soviet air base in Estonia: 'Complete closure is the cheapest option, but this means abandoning re-use plans and would involve large expense in cleaning up the degraded environment'.

Land use and attitudes

Any human activity requires a portion of land in the quantitative sense of surface; agricultural activities need land also in the qualitative sense. Land, however, is a limited resource: only 29% of the globe's surface lie above sea-level, and only part of this surface can be put to some use by man. Thus, about 24% of the total land area could be cultivated, but only about 11% are actually used for agricultural purposes (Pierce, 1990, p.28). Environmental limitations (partly natural, partly due to human interference with the ecosystem) affect about 90% of all soils (*ibid.*, p.21).

If these statements are a commonplace, they are nevertheless of primary importance. Looking at the way humanity deals with resources, it seems as if people had forgotten such simple facts. Rationally speaking, land (just as any other non-renewable resource) ought to be used sparingly, taking into account quantitative and qualitative aspects and judging wisely the kind of land use which would be most appropriate for a given society. This fact is emphasised by planning laws. The European Planning Charter of 1983 demands a rational use of land (article 17), and the Swiss Planning Law of 1979 requires the Confederation, the cantons and the communes to use land moderately (article 1). But how do we interpret 'rational use' or 'moderately'? Such norms are subjective, guided by the worldview shared by the majority of people (or rather by the primary economic and political actors and decision-makers).

Land use is located at the interface of socio-economic and bio-physical systems, i.e. between subjective needs formulated by humans and objective constraints due to environmental conditions (Figure 2.1). In theory, land use ought to be in a balance between those two.

What has been happening over time, however, is a process of reckless exploitation, which manifests itself in numerous ways.

- Agricultural monocultures exhaust the soils and require an increasing input of artificial fertilisers and pesticides; the law of diminishing returns is not taken into consideration.
- Fossil resources are being extracted as if the reserves were unlimited.
- The base for the survival of renewable resources is being destroyed by excessive exploitation (e.g. deforestation, overfishing).
- In the 'one-way-society' (or 'throw-away-society'), the durability of consumer goods has been replaced by constant innovation; in many cases the new products are not really new but contain just minor adjustments according to fashion. However, additional waste is created in this way.

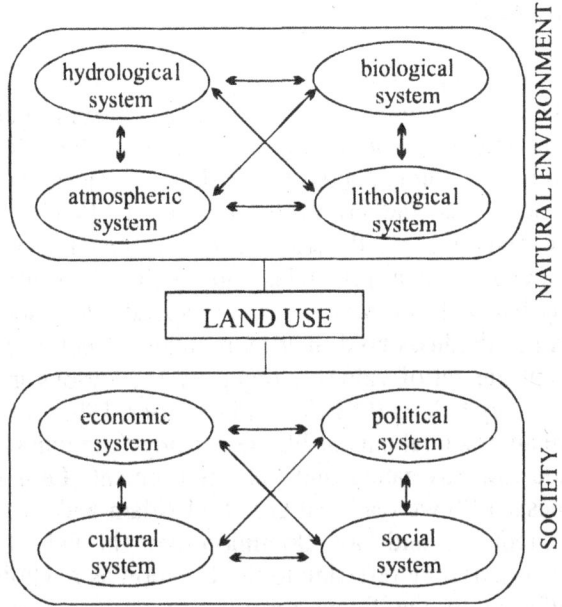

Figure 2.1 A systemic approach to land use

These four examples demonstrate that input and output reveal themselves as dangerous for both society and environment: remnants of toxic substances gradually enter the food chain and the water cycle, soil erosion and degradation as a consequence of deforestation reduce the surface available for forests and farming, and superfluous products as well as waste from the production process are dumped somewhere in the landscape. There is a lethal feedback, a 'boomerang-effect' (Beck, 1986, p.48): it hits back upon those who are at the origin of this input. In Shakespeare's words: '...; that we but teach bloody instructions, which, being taught, return to plague the inventor' (Macbeth I.vii.).

Man himself is responsible for his actions which are taken following a male world-view (male and female as two principles, not the gender of a person, which coexist in every person and represent a dynamic equilibrium; Capra, 1986, p.32); he may reap the benefits and has to bear the consequences. Behind every decision, there is a set of norms or values which guide us and on which we rely in every moment of life. Our attitudes and our perceptions of the world are coined by them, and we do not like to change them. Although life is dynamic, human attitudes are characterised by inertia – another commonplace to which everybody can testify.

Values and world view

Human existence is unimaginable without a value system. Such a system can be described as a complex of concepts, which guide human thinking and actions, serving as a sort of fundamental guideline. Values are the elementary motives of human activity, the desired objectives or end-states of a process, standards and criteria of orientation (Weber, 1980, p.12) concerning the behaviour of the members of a social group (Hillmann, 1989, p.53), they are general principles legitimating the rules of behaviour set by norms (Chazel, 1988, p.125), and are specific to a society (Beattie, 1964, p.73). All individuals hold their own value system, even if they will ultimately accept the set of values adopted by the various groups and the society to which they belong (O'Brien and Guerrier, 1995, p.xiii).

Values are an essential part of life, governing both our social relations and our attitude towards nature and the environment (Leimgruber, 1995, p.393). In Genesis 1/28 we read that God told Adam and Eve to 'replenish the earth and subdue it: and have dominion over the fish of the sea' etc. With this, He was certainly referring to the Creation as a whole, even if He specifically referred to the living creatures, i.e. the biotic part of the Creation. Throughout history, the notions of 'subduing' and of 'having dominion over' have been understood as 'management'; however, since John Locke, the European civilisation has gradually interpreted them as 'exploitation'. Very probably, God meant the former rather than the latter, at any rate never 'deterioration' or even 'destruction'. Even if we can understand Locke's argument, i.e. that subduing nature is a means to avoid scarcity (Van Dieren, 1995, p.19), this can by no means serve as an excuse to what has happened to the Creation since.

Exploitation and management mirror two extreme values which can be imagined as lying at the ends of a continuum which can be called the 'value system' (Hillmann, 1989, p.141; Figure 2.2). Every individual, every group and every society can be located somewhere along this continuum, and this position is not stable over time.

Figure 2.2   The value system

The two extremes can be described as follows (after Hillmann, 1989, p.142):

- *Secular values* are open towards innovation and the adaptation of new and outside ideas. They are dynamic, forward-oriented, tolerant, rational, emphasising modernism, change, functional relationships, uniformity, individuality and competition, and they include the notion of exploitation. Secular values correspond to the male or *yang* view.
- *Sacred values*, on the other hand, are characterised by stability, persistence, conservation, tradition, diversity, community and solidarity, an emphasis on security and distrust of innovation, and their outlook is retrospective. Actions are likely to be guided by emotional considerations rather than by rational ones, and they favour management and sustainability. Sacred values correspond to the female or *yin* view.

It would be wrong to qualify either of these two extremes as positive or negative. Every society and every human being needs elements of both of them in order to survive. Privileging the secular side will culminate in the fragmentation and dissolution of a society, whereas the exclusive dependence on sacred values will result in an extremely static society, unable to take up the challenges, which are incessantly facing it. Interplay (a dynamic equilibrium) between the two is therefore essential, but it is difficult to find the right blend, which can assure a harmonious development. Sacred values, however, can be considered as primary or supreme because they constitute the framework within which secular values will become operative.

This statement is confirmed by Redfield (1968, p.25) who discusses the elements binding early societies (folk societies) together and uses the notions of 'technical' and 'moral' order. 'The 'moral order' refers to the organisation of human sentiments into judgements as to what is right'. (*ibid.*, p.32), whereas 'in the technical order, men are bound by things, or are themselves things. They are organised by necessity or expediency'. (*ibid.*, p.34). In primitive (or folk) society, the two order systems were clearly defined: the moral order dominated and the technical order was subordinate to it (*ibid.*, p.30). In modern society (civilisation), the relationship between them has become more complex, even unstable: the balance is precarious, at times one could believe that the technical order had taken over, but then the moral order comes back again.

Throughout modern history, the European society has gradually shifted away from sacred to secular values, privileging attitudes such as individualism, efficiency and profit. The result has been an increase in production and material wealth, but a loss of respect for nature and environment or for the weaker part of the Creation as a whole: individual well-being dominates over the common good. The relations of man to the land have thus been governed by the technical order, and territorial development has aimed at obtaining a maximum profit at no cost: the environment is priceless, it is omnipresent (a ubiquity) and seems to be waiting to be exploited. Both the free market and the socialist economy never included environmental considerations into their calculations, neither in their national accounts nor in production cost. Taking full account of environmental costs would result in higher prices for consumer goods (as is the case with organically produced food or with the recycling-tax levied on electronic goods and batteries, for example) and in a reduced or even negative growth of national economies (Van Dieren, 1995, p.74). Economists begin to acknowledge that these costs arise 'from unintended effects of production and consumption' (*ibid.*, p.75).

The shift from the dominance of sacred to that of secular values that characterised European society is tied to our changing worldview. Lea (1994) has shown how the relationship between man and God has evolved since the Middle Ages: once, God dominated His Creation, now He has gradually been replaced by man and his own creations (Figure 2.3).

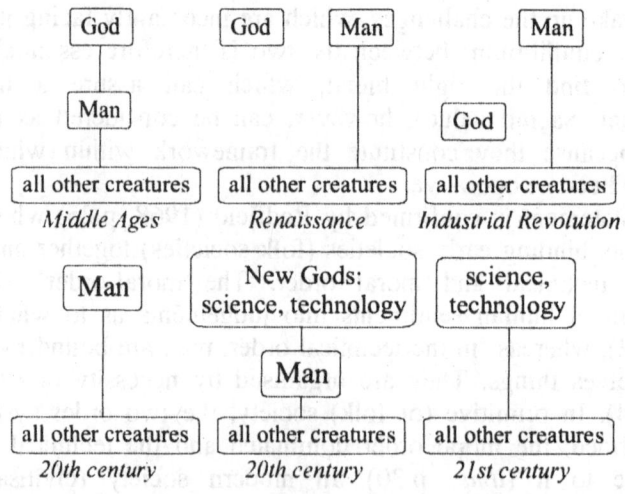

Figure 2.3   World views since the Middle Ages

If in the mediaeval period, God was represented as the One governing over mankind and the whole Creation, Renaissance thinking has led man to elevate himself to a position almost equal to God. Later, the industrial revolution demonstrated the superiority of human creativity, and the 'old God' was gradually forgotten: man became the sole dominator of Creation. Finally, 'new Gods' (such as technology, money and finance, eternal health) replaced the 'old God', dominating human thinking and activities, and they may well take over to the extent that humans can be disposed of.

Geography and religion – religion in geography

The evolution sketched above invites us to reflect on the relations between geography and religion. Scientifically, they represent two disciplines lying wide apart and belonging to different university faculties. However, spiritually speaking they are very close to one another: 'If we seek after the nature of God, we must consider the nature of man and the earth, and if we look at the earth, questions of divine purpose in its creation and of the role of mankind inevitably arise'. (Glacken, 1967, p.35). The existence and acceptance of God is the prerequisite to any religion, and every religion is tied to the definition of God. In a very broad perspective, He can be seen as Spirit, as the General Principle of the Universe, as the sum of the Spiritual Laws, as the Party, as Capital, as Sports or as anything else. It is true that this is a rather unconventional way to understand religion, but the elements of belief and institutions are present with the Party, Capital, Sports etc., and often such 'new Gods' are considered a 'Supreme Being whose authority is distinct from worldly powers'. (Gottlieb, 1996, p.8). Religion can therefore be seen as an equivalent to the value system. Whether we believe in God as transcendental or raise, e.g., sports to the level of God – there is always a guiding principle for our life. Tuan (1976, quoted in Park, 1994, p.34) comes close to this when he calls religion 'the impulse for coherence and meaning' (*ibid.*, p.271). Linguistically speaking, religion means fear of god, piety, but it could also be understood as binding (back to God). Whatever its true meaning, it is primarily spiritual or irrational. It defies rational scientific study, but it is nevertheless fundamental to mankind.

Of capital importance in this discourse on religion are the distinction between its external (or human) and its internal sides. On the one hand, religion comprises *human institutions*, which take care of concrete aspects such as the practising of worship, the care for the adhering group (congregation, community) and for the society as a whole. This is what according to Saint Augustine (De Civitate Dei I, p.57) can be called the visible or external Church. On the other hand, religion is the *teachings* of prophets (Jesus Christ, Mohammed, Confucius, Buddha etc.) whose intention was to furnish moral guidelines to human life – according to Saint Augustine (*ibid.* I, p.57) the invisible or internal Church. These teachings are of long-lasting value, whereas institutions are limited to human life expectancy and man's changing interpretation of the way institutions should be run and organised. Unfortunately, every society tends to interpret the rules laid down by the founders according to its own ideas, hence the discrepancy between the ideal and reality, between teachings and actions (Park, 1994, p.6).

The interactions between geography and religion have to take these two fundamentally different contents of the term religion into account. Unfortunately, in most cases the institutional aspects dominate the scene. This is demonstrated by the list of topics, which Park (1994, p.2-7) proposes for his book on geography and religion:

- Geographical distribution of religions
- Religious imprint on the cultural landscape
- Impacts of religion on lifestyle and commerce
- Religious taboos on food and wildlife
- Religion and demography
- Religion, politics and conflict
- Religion and culture
- Religion and environmental attitudes

The emphasis is clearly on the 'external Church', only the last point can be attributed to the 'internal Church' – but precisely this theme is not treated in his book! Man's attitudes towards land guide the use he makes of it, and attitudes are based on a society's value system: exploitation and profit, or respect and sustainability.

Traditional societies based their relationship to land on religious arguments. Being non-rational, religion evokes 'emotion in moral codes' (Anderson 1996, p.161), and this moral code has to be believed by the people who must be convinced by the rightness of this code. A forest inhabited by spirits, a pond where a nymph dwelt – they were protected from abusive human action because of the sacred nature of the 'inhabitants'. It does not matter whether the people believed in the real existence of these spirits or nymphs; the result (i.e. respect for the forest, for the pond) was important.

Sacred land – marginalized land

Every society has a specific relationship to land, which grounds on a spiritual conviction. Land use is motivated not only by materialistic but also by spiritual considerations, as Cohen (1993, p.5) notes in his discussion of Israeli – Palestinian relations: 'the meaning of land and the practices relating to it ... have religious, historical, social, economic and political significance'. All relationship to land is religious, however, the kind of religion (and of God) has to be defined. Taken in the more restricted sense of the teaching of moral guidelines beyond the material world, religion should inspire a sense of respect to land, to the environment, to the Creation. This kind of relationship is strongly rooted in folk societies. The American Indians, for example, practised a variety of land uses, they hunted and fished, but their activities exercised little negative impact on the environment: 'The land was not untouched, but it was unspoiled'. (Hughes, 1996, p.132). The clash between the Indian culture and the white civilisation was also a clash between two different mentalities, two ways of interpreting man's relationship to the land. The whites acted according to their interpretation of the Genesis (1,28) and viewed the 'empty' land as a territory to be subdued. The Indians' lives, on the other hand, were guided by 'ecological wisdom' based on the belief that all manifestations of nature were powerful spirits (*ibid.*, p.133). Animals, trees, mountains etc. were sacred, i.e. required respect. This holds good even if John Locke did not see it this way. To him, the American Indians were 'rich in land, but poor in all the comforts of life' (van Dieren, 1995, p.21) – a statement which justified colonisation by the industrious Whites. Of course, his assertion grounds on a specific understanding of 'comforts of life' and a loss of religious feeling. The secularised Europeans who discovered and conquered the land had already lost the spiritual relationship to land – it was there, seemingly empty, hence it was to be occupied and exploited.

This same attitude still dominates in certain parts of the world. The policy of transmigration in Indonesia provides an example of how little attention is paid by the central government in distant Java to the ways of life of indigenous populations on remote islands. The crop rotation practised by many indigenous tribes requires certain areas to lie fallow; idle land, however, is considered, as unoccupied and unoccupied land is state property. Such surfaces will therefore be allocated to transmigrants, to the detriment of the original owners. Not only do they lose the material base of their existence, they are also bereft of the symbolic or mythic character of their land: 'To rob an ethnical group of its territory means to deprive it of part of its self'. (Seydoux, 1998). The sacred aspect of land has to yield to secular political considerations.

A particularly striking example of changing values is furnished by post-World War II Finland. Under the Land Acquisition Act of 1945, some 2.6 million hectares of land had been allocated to evacuees from Karelia and war veterans (Varjo, 1977, p.26). Colonisation took place all over the country, including the north, where 'fields usually had to be cleared from forests and peatlands' (Rikkinen, 1992, p.52). The Land Utilisation Act of 1958 enabled another 1.3 million hectares to be settled and cultivated. By 1969, about 135,000 new farms had been crated. However, since the 1950s, agricultural production had begun to exceed domestic consumption, and a cut-down in production was called for. In 1969, a field reservation law was passed which offered farmers a premium per hectare field area laid fallow and per head of cattle slaughtered. By 1973, some 230,000 hectares of field area had been taken out of production (Varjo, 1977, p.28), a figure which was to rise to almost 300,000 hectares (about 10% of the total field area) until 1989 when the scheme was discontinued (Häkkilä, 1991, p.45). The field reservation law did not reduce overproduction much, as mainly small farms participated. On the other hand, its social consequences were far reaching, because it contributed significantly to the general structural change of agriculture: farmers felt encouraged to retire from business, and the young generation migrated to urban areas (Varjo, 1977, p.29). Part of the abandoned land reverted to natural vegetation, part of it was afforested.

On the other hand, land abandoned as a consequence of rural out migration is a characteristic in many parts of the Alps. Rural – urban migration has been particularly vigorous after World War II, resulting in an ageing of the resident population and a gradual withdrawal from farming activities. The small size of the land plots, due to the former practice of the division of holdings, favoured this process. Most land has reverted to natural vegetation – the positive aspect is that the forest area in Switzerland has increased from 25% in 1950 to about 30% in 1990.

The disposal of refuse is probably one of the best examples of the marginalization of land. The waste output of the economic system requires a certain surface just as production plants, transport systems and dwellings do. Whereas the latter are eagerly promoted, the disposal of refuse is a different story. Pre-industrial societies were not confronted with this problem because all waste was returned to and absorbed by nature. From the industrial revolution onwards, however, we were confronted with the dilemma what to do with that unwanted output which would not be recycled by the ecosystem. Waste was considered as superfluous material, which results from the production process and from households. It had to be eliminated as quickly and as cheaply as possible – preferably by dumping it in some remote place (small valleys, forest clearings, old quarries and the like) considered of no economic value and therefore marginal in people's perception. The growth of cities and the alienation or humans from natural processes has accelerated this attitude.

To set the balance right: waste is not simply waste. In certain circumstances, it may be an interesting and even important material, demonstrating that our northern perspective is not the only one possible.

- Archaeologists collect a wealth of information about past living and working habits from waste deposits.
- To thousands of people in urban peripheries of Africa and Latin America, waste dumps are of central importance: as long as waste is being dumped, something can still be got out of it. However, once incineration will become the general practice, this kind of precarious activity (marginal economy) will cease to exist.
- In Ouahigouya (Burkina Faso), Laurent (1994, quoted in Leimgruber 1995, p.403) found that domestic waste dumped in collective pits consisted to 80% of gravel, sand and earth (from sweeping the courtyards), a further 18% was organic material (most of which would otherwise be used for compost), but that there was no metal waste.

The examples quoted above show that it has become difficult to maintain the sacred aspect of land in the restricted (i.e. non-materialistic) sense of holiness applied in this chapter. We have lost the respect for land, as a consequence we are striving to find a new and balanced relationship to it. This may in part be the result of alienation: most of our modern activities are no longer dependent on land as a resource for our livelihood but simply as a surface for the infrastructure required. The perception of land is different, if we live with it or if we consider it as a resource at our service. 'This perception of national economy which the state today represents is not the perception of the people for whom forests, land, and water together comprise a primary life-support system'. (Sharma, 1996, p.558). Under such condition, one is more easily ready to call land 'a sacred trust of humankind' (*ibid.*, p.559). This implies a trustful relationship to God, as Pope John Paul II evoked in a message in 1989. As a model he mentions Saint Francis of Assisi who 'gives us striking witness that when we are at peace with God we are better able to devote ourselves to building up that peace with all creation which is inseparable from peace among all peoples'. (Pope John Paul II, 1996, p.236).

The Pope's message points to the close relationship between the problems of the environment (degradation, pollution) and those of the society (crime, drugs). It also emphasises the fact that God is not simply an abstract (and non-existent) being but can be encountered everywhere in His Creation, i.e. in the land as well as in the people.

This leads us to a further point concerning marginalization. Any study of marginal regions has to do with three aspects: perception, scale and time.

'Marginality is a state of mind' means that its definition depends on people's perception. In order to recognise marginality, we require a point of reference upon which we can compare an existing situation. This reference can be income, job or leisure opportunities, schools or anything else. The evaluation and eventual definition of marginality depends on the values of the society concerned. A person earning 1000 US$ a month will belong to a marginal group in Switzerland but to a privileged (central) one in – let's say – Burkina Faso.

The scale component has been amply demonstrated by the CLUE-Group of Wageningen Agricultural University: 'In agro-ecosystems analysis, the scale at which the analysis is conducted will affect the type of explanation given to the observed phenomena. While pests and diseases might cause variation within a rice field, climate systems determine broad agro-ecological zones. At coarse (aggregated) scales, the high level of aggregation of data obscures the local variability but can show patterns invisible at detailed scales, and vice versa. Furthermore, factors determining land use (change) can operate at great spatial distance from the area affected. Thus, for dealing with the complex issues of land use/cover change, it is necessary to use a multi-scale approach that identifies and quantifies land use drivers and their interrelationships at various spatial scales'. (De Koning *et al.*, n.d.). What may appear, as marginal at a large scale will gradually disappear if we change to ever-smaller scales. A local ecosystem may suffer from local pollution, but the regional system may remain largely unaffected. A small waste dump will cause disturbance to its immediate neighbourhood but be of little avail to a large region.

Ecosystems are in constant flux, nothing is stable, and things change over time. What has been central today may be marginal tomorrow and become central again thereafter. Early industrial areas, such as Northeast England, Wallonia or the Ruhr used to be of central importance to national economies – until the crisis in the steel industry, resource depletion, new industrial orientations etc. put an end to this position: they became regions heavily dependent on regional development policies.

Marginality can also be considered from an ecological perspective. In this case we have to modify our anthropocentric and economically minded understanding of marginality and centrality (Leimgruber, 1994). Traditionally, human activities and artefacts are considered as superior to nature, and nature as an enemy to mankind (after 16th century interpretation, Van Dieren, 1995, p.17; Figure 2.4).

Figure 2.4 Traditional way of viewing marginality

However, if we consider nature as the essential basis for human existence, nature and the natural landscape will become central, whereas culture or civilisation and the cultural landscape become marginal (Figure 2.5).

Figure 2.5 A new way of viewing marginality

Nature and culture lie along a continuum. 'From this continuum it can be seen that increasing human activity will marginalize the environment. Since this environment includes the fundamentals of human existence, this kind of marginalization leads to the degradation of this basis'. (Blötzer *et al.*, 1997). It follows that no human culture can exist without favourable natural conditions, that an ecological marginalization will result in a decline of the quality of life, i.e. have socio-economic and socio-cultural consequences. Human activities have therefore to be based on the principle of sustainability, which takes into account that 'all species interact, change and co-evolve with their environment. The human species is no exception. We are exceptional, however, in our ability to modify consciously some elements of the pattern of our interaction with the environment'. (Clayton and Radcliffe, 1996, p.7). Sustainability would therefore mean that there is an equilibrium between the social and the ecological system, based essentially on sacred values.

Concluding remarks

Land use has existed ever since the advent of the human race. Throughout history, man was conscious of the importance of land, nature and environment to his activities and his survival. Land use decisions were and are still guided by the prevailing worldview of a society, a view which was originally directed towards a harmonious (sustainable) coexistence with the ecosystem, but which is now dominated by technical considerations. It is true that the problems our society faces nowadays urge us to look for more technical solutions, but every technical success hides a problem: nuclear energy promised the solution to the shortage of electric energy and a way out of atmospheric pollution; concentrated animal feed, chemical fertilisers, pesticides etc. seemed to provide the solution to food shortage; genetic engineering promises another solution to food shortage plus improved health and long lives. And then we lived through the nightmares of Chernobyl and BSE, we watch the degradation of soils – and all our confidence in modern technology is shattered. It has become obvious that despite considerable progress, technical solutions alone cannot solve our problems.

A new world view is required which leads us back to the moral order, to a consciousness of what is right for humanity. A shift back on the value continuum from secular towards sacred values is necessary. Such a worldview might run as follows:

> The biblical saying 'subdue the earth' is not to be understood from the point of view of a power-stricken ruler but of the caring gardener. We are material beings (bodies), equipped with a soul and a spirit, which need non-material 'food'. It is therefore imperative to use the resources of the ecosystem earth with care. The current problems of both environment and of society mirror greediness, hedonism and a lack of reason.

Nursing the ecosystem means not to stretch its limits too far. It is true that the limits of the ecosystem are 'elastic', but every rubber band will break at a particular moment and so will the ecosystem.

While this all sounds very pessimistic, there is still some ground for optimism. After all, the destiny is not cruel but consistent (Wolfgang Döbereiner, pers. comm.) a consistency already known to our elders (cf. the quotation from Macbeth). Indeed, the first results of a new thinking are already visible, as can be seen by numerous examples of grassroots or bottom-up movements all over the world (Ekins, 1992). Two examples from Switzerland may briefly illustrate this shift:

1. Since about 1990, organic farming has known an unprecedented success all over Europe including Switzerland (Leimgruber et al. 1997; Table 2). In part, this stems from an increasing consumer awareness, in part also from a growing concern of the farmers – the interplay between demand side and supply side is an important element of this process, and retail trade has quickly reacted to it. Besides, however, agricultural policy has had its share in the shift from conventional to organic, but all measures, which are to succeed, need to be supported by the political level. Politicians have to become aware of the problem and take the measures, which will help to solve it. In the case of organic farming, it was not an intentional move towards sacred values; the new agricultural policy aims at the preservation of the environment because the alarm bell has been sounded. Ecological transfer payments are an apt instrument to induce farmers to change their philosophy and operate in a sustainable way. The results of the past few years are impressing (Table 2.1), even if only a fraction of the farmland has so far been converted: the surface devoted to organic farming has increased from 0.9% of the agricultural land to 6.6% within seven years – a figure and a tendency below that of Austria (from 0.5% to 8.7% in the same period) and of Liechtenstein (18% of total agricultural area in 1997), but well above that of other European countries.

Table 2.1   Organic farming in Switzerland in the 1990s

	1990	1993	1996	1997
Number of Farms	1,031	1,386	3,721	4,278
Farming area (ha)	10,079	20,784	59,675	71,790
Cereals (ha)	1,233	1,320	2,863	3,426
Fruit trees (ha)	90	60	173	232
Vegetables (ha)	167	280	473	800
Grassland (ha)	8,589	19,124	56,166	67,332

Source: Research Institut on Organic Farming, Frick.

2. The regional dimension has been rediscovered in a variety of economic and cultural domains (Leimgruber and Imhof, 1998). Mass tourism and mass importation of exotic food is countered by regional initiatives, promoting soft or sustainable tourism, and the regional marketing of regional products either through shops or restaurants or through direct sales by farmers. Classical music festivals in remote areas meet with an increasing success (e.g., in Ernen, a small village situated off the main road in the Upper Valais) – the local and regional scale close to the people and to nature has

become a new attraction for the urban population living under a permanent social and environmental stress. Again, various interests meet: people from the cities seek refuge in the mountains, and their demand promotes local initiatives, combined with regional and cantonal policies. Landscape protection in this case is paramount, because the risk of natural catastrophes is always present. This is emphasised by Stone (1992, p.68) who suggests that the variety of alpine ecosystems has to be respected, 'specific land use practices must be chosen not only to produce food but also to ensure ecological stability', and that certain slopes be not cultivated at all. In addition, 'a great deal of maintenance and repair work will also be necessary to give cultivated land additional stability'.

These two examples demonstrate that the pendulum is gradually swinging back from secular to sacred values, from technical to moral order – the sacred is creeping back silently but efficiently. Nobody speaks of religion or God in this context, but they are nevertheless present. After all, 'every society needs ... sacred places. They help to instil a sense of social cohesion in the people and remind them of the passage of the generations that have brought them to the present. A society that cannot remember its past and honour it is in peril of losing its soul' (Indian wisdom by *Vine Deloria jr.*, quoted in Birdsall, 1996, pp.626).

References

Anderson E.N. (1996), *Ecologies of the heart. Emotion, belief, and the environment*, Oxford UP, New York.
Beattie J. (1964), *Other cultures. Aims, methods and achievements in social anthropology*, Routledge, London.
Beck U. (1986), *Risikogesellschaft. Auf dem Weg in eine andere Moderne*, Suhrkamp, Frankfurt/M.
Birdsall S.S. (1996), 'Regard, respect, and responsibility, sketches for a moral geography of the everyday', *Annals of the Association of American Geographers* 86/4, pp.619-629.
Blötzer A., Keller F., Piller F. and Portmann D. (1997), *Landnutzungswandel und Randregionen: das Beispiel des Aralsees*. Paper prepared during a Seminar on marginal regions, Department of Geography, University of Fribourg, Manuscript, Fribourg.
Capra F. (1986), *Wendezeit. Bausteine für ein neues Weltbild*, 11.A, Scherz, Bern.
Chang-yi D.Chang, Sue-ching Sue and Yin-yuh Lu (eds.) (1994), *Marginality and Development Issues in Marginal Regions*, Proceedings of the Study Group on Development Issues in Marginal Regions, Taiwan 1 – 7 August 1993, National Taiwan University, Taipei.

Chazel F. (1988), 'Normes et valeurs sociales', *Encyclopédia Universalis*, vol. 13, pp.124-127

Clayton A.M.H. and Radcliffe N.J (1996), *Sustainability. A systems approach*, Earth-scan, London.

Cohen S.E. (1993), 'The politics of planting. Israeli-Palestinian competition for control of land in the Jerusalem periphery', *University of Chicago Geography Research Paper* Nr. 236.

De Koning G.H.J., Veldkamp A., Verburg P.H., Kok K. and Bergsma A.R. (n.d.), *CLUE: A tool for spatially explicit and scale sensitive exploration of land use changes*, Internet http://www.gis.wau.nl/~landuse1/lima.html (07.07.1998)

Ekins P. (1992), *A new world order. Grassroots movements for global change*, Routledge, London.

Glacken C.J. (1967), *Traces on the Rhodian Shore*, UC Press, Berkeley.

Gottlieb R.S. (ed.) (1996), *This sacred earth. Religion, nature, environment*, Routledge, New York.

Guerrier Y., Alexander N., Chase J. and O'Brien M. (eds.) (1995), *Values ad the environment. A social science perspective*, John Wiley, Chichester.

Häkkilä M. (1991), 'Some regional trends in Finnish farming with special reference to agricultural policy', *Fennia* 169:1, pp. 39-56.

Hillmann, K.-H. (1989), *Wertwandel. Zur Frage soziokultureller Voraussetzungen alternativer Lebensformen*, Wiss. Buchgesellschaft, Darmstadt.

Hughes J.D. (1996), 'American Indian ecology (extract)', in R.S. Gottlieb (ed.), *This sacred earth. Religion, nature, environment*, Routledge, New York, pp.131-146.

Jauhiainen J. (1997), 'Militarisation, demilitarisation and re-use of military areas', *Geography* 82/2, pp.118-126.

Jauhiainen J. and Lootus H. (1998), 'Entiset sotilasalueet Virossa: ympäristöriskit ja uuskäytön mahdollisuudet', *Terra* 110/1, pp. 3-11 (Former military areas in Estonia: environmental hazards and possibilities of re-use).

Lea D.R. (1994), 'Christianity and western attitudes towards the natural environment', *History of European Ideas*, vol. 18, No. 4, pp.513-524.

Leimgruber W. (1994), 'Marginality and marginal regions: problems of definition' in, Chang-yi D.Chang, Sue-ching Sue and Yin-yuh Lu (eds. 1994), *Marginality and Development Issues in Marginal Regions*, Proceedings of the Study Group on Development Issues in Marginal Regions, Taiwan 1 – 7 August 1993, National Taiwan University, Taipei, pp.1-18.

Leimgruber W. (1995), 'Local efforts and people participation in the development of marginal regions. A summary of 10 years' research at the Fribourg Department of Geography', *Boletín de Estudios Geograficos, Universidad Nacional de Cuyo, Facultad de Filosofía y Letras, Instituto de Geografia*, 26, pp.391-408.

Leimgruber W., Dettli W. and Meusy G. (1997), 'Organic farming: a solution to marginality?', *Proceedings of the Symposium 'Issues of environmental, economic and social*

stability in the development of marginal regions: practices and evaluation', Glasgow, pp. 160-169.
Leimgruber W. and Imhof G. (1998), *Remote alpine valleys and the problem of sustainability,* International seminar 'Exploring sustainability: on the future of small society in a dynamic economy', Karlstad (Sweden), 12.-14.05.
Mather A.S. (1986), *Land use,* Longman, London and New York.
O'Brien M. and Guerrier Y. (1995), 'Values and the environment: an introduction', in Y. Guerrier, N. Alexander, J. Chase and M. O'Brien (eds.), *Values ad the environment. A social science perspective,* John Wiley, Chichester, pp.xiii-xvii.
Park C.C. (1994), *Sacred worlds. An introduction to geography and religion,* Routledge, London.
Pope John Paul II (1996), The ecological crisis: a common responsibility. in R.S. Gottlieb (ed.), *This sacred earth. Religion, nature, environment,* Routledge, New York, pp.230-237.
Redfield R. (1968), *The primitive world and its transformations,* Penguin, Harmondsworth.
Rikkinen K. (1992), *A geography of Finland,* University of Helsinki, Lahti Research and Training Centre.
Seydoux A. (1998), 'La transmigration en Indonésie et ses impacts sur les régions périphériques', Paper prepared during a Seminar on marginal regions, Department of Geography, University of Fribourg, Manuscript, Fribourg.
Sharma B.D. (1996), 'On sustainability', in R.S. Gottlieb (ed.), *This sacred earth. Religion, nature, environment,* Routledge, New York, pp.558-564.
Simmons I.G. (1989), 'Changing the face of the earth, Culture, environment, history', Blackwell, Oxford.
Stone P.B. (1992), *The state of the world's mountains, A global report,* Zed Books, London and New Jersey.
Van Dieren W. (ed.) (1995), *Taking nature into account, A report to the Club of Rome,* Springer, New York.
Varjo U. (1977), *Finnish farming,* Geography of world agriculture 6, Akadémiai Kiadó,: Budapest.
Weber M. (1980), *Wirtschaft und Gesellschaft,* 5th ed. Mohr, Tübingen.

3 Trajectories of change in marginal and critical regions in Southeast Asia and Southern China

R.D. HILL

Introduction

Marginal and critical parts of the region have already seen some degree of economic integration into national and global economies and thus are affected by the processes of globalisation and structural change. Using a time frame of about 30 years a number of trajectories of change are hypothesised. These include a continuance of the status quo, though in the short term as a result of current economic difficulties perhaps some reduction in the impact of 'outside' forces may be experienced. It is argued that temporary out-migration is a 'safety-valve' in the face of continued population growth, permitting the maintenance of a status quo in which people and resources are more or less in balance. The second trajectory is a downward spiral initiated by soil erosion on intensively cultivated sloping lands. A further, familiar downward spiral is that characterised by decreased forest or scrub fallows, over-exploitation of forest in the face of continued population growth and few employment opportunities outside the local economy. Major responses are discussed – the shift to perennial tree-crops, or in limited areas, teak, to intensive cropping, and to the adoption of opium. The article concludes by identifying a possible 'upward spiral' that would include tourism and other activities controlled by marginal peoples themselves. This outcome is likely to require affirmative governmental action and access capital, skills and other resources not likely to be available to most in the foreseeable future.

Southeast Asia and southern China

There can be little doubt that Asia in general and Southeast Asia and southern China in particular are for the most part participating in major economic and social change – quite apart from the recent and in some cases, severe economic downturn. Although the maintenance of some degree of local control over the subsistence base has to some degree cushioned marginal regions from the effects of the economic slowdown, such is degree of their economic integration that few have remained totally unaffected.

The region comprises three broad swathes of land extending from west to east. The southern-most includes Indonesia, Malaysia and the Philippines, broken up into large and small islands, some mountainous – Sumatra with the Barisan Range, Borneo with the Crocker Range (including the Kinabalu Massif), Java with its central volcanic core, Luzon with the Cordillera Central. These major lineaments are flanked by extensive tracts of hill country of moderate elevation but usually steep slopes. The second belt is made up of mainland Southeast Asia from central Burma to Viet Nam. This region is also hilly, becoming mountainous especially northern and eastern Burma where a continuation of the Hindu Kush-Himalaya massif curves southwards in an area of north-south trending ranges paralleled by rectilinear gorges cut by major rivers such as the Irrawaddy, Salween and Mekong incised to a depth of 1000 to 2000m. In their lower courses these and other rivers such as the Chao Phrya and the Hung, broaden to form wide alluvial plains intensively used for rice. The third swathe includes northern Burma, margining the Tibetan Plateau at almost 6000m, but mostly comprising a set of gigantic topographical 'steps' from about 3000m in Yunnan and Guizhou becoming progressively lower eastwards through Guangxi and Guangdong but rising again to around 1500m in Fujian.

Climatically the region is monsoonal being dominated by cool, dry air from the east or north-east in winter, except in the equatorial zone where the same dominant airmasses, in crossing warm seas, are heated and moistened as a consequence. In summer the region comes under the influence of the southerly monsoon, southwesterly or southerly in mainland Southeast Asia and southern China, southerly to easterly in insular Southeast Asia. Except in the eastern Indonesian Archipelago this is almost universally the season of heavy precipitation though the intermonsoon periods may also bring much. Lands fringing the South China Sea are also wet in winter. Annual totals range from about 1m to 5m (less in the double rain-shadow area of Burma's Dry Zone), characteristically falling in episodes of 2-4 days of high-intensity rain, or as short, torrential convective showers where and when the monsoonal circulation is weak.

Soils are deeply weathered except on the steepest slopes, though the depth of weathering broadly decreases with increasing latitude and elevation. They are mostly easily erodible and, given the rainfall regime, become particularly so where soil parent materials are weakly consolidated – recent volcanics of Java, Bali and parts of the Philippines – and in seasonally-wet regions of savannah vegetation. Cultivation on slopes results in rates of surface erosion three to four orders of magnitude higher than under forest whereas perennial tree-crop cultivation increases them by about one order of magnitude (Hill and Peart, in press).

Within the region it is not possible to define areas which may be considered marginal, not least because there are degrees of marginality. The total population is about 715 million, of which 295 million live in insular Southeast Asia, some 180 million in mainland Southeast Asia from central and southern Burma to Viet Nam, and a further 240 million in northern Burma and the southern Chinese provinces (excluding Hainan). At a guess, about a tenth may be considered to be living in upland marginal areas, leaving aside such stable agricultural areas as much of Java, Bali, parts of north-central Luzon, Yunnan, Guizhou, where terraced wet rice production allows dense populations to be supported in a sustainable manner. Seventy million people are by no means negligible, being roughly equivalent to the total population of the Philippines, Viet Nam or Guangdong Province.

Marginal regions are largely occupied by minority peoples in bewildering variety, often spreading across international boundaries in upland regions: Shan (T'ai) in Burma, Wa also in that country and in China, Miao in China, Viet Nam, Thailand. Laos alone contains 40-60 ethnic groups and the region as a whole has literally hundreds of other groups ranging in size from a few hundred to a few million. Many are shifting cultivators, some with permanent rice-fields, some without. Some are hunters and gatherers though not solely for subsistence since various jungle products, the raw material for chewing gum is one, rattan is another, have long been traded with outsiders. Others cultivate sloping lands on a permanent basis, especially in NE Thailand and parts of Southwest China.

While there may be groups in remotest northern Burma, a country itself far from fully integrated into the world community, in parts of Laos, southwestern China and, especially, Indonesia's Irian Jaya, whose links with the outside world are tenuous and indirect, these are clearly exceptions. If they have not yet felt the impacts of structural change in regional and local economies, it can be confidently asserted that they soon will even if they attempt to isolate themselves by moving further into jungles, themselves under threat. At the same time globalisation continues to reach into remote areas, whether under the guise of eco-tourism, as dam-building or in the form of education and health care or the imperatives of national security.

Structural change has seen a massive shift out of agriculture and into manufacturing and the service sector. But this shift has only been relative. In every developing country in the region, almost certainly including southern China (though specific data are lacking), there are now more agriculturists than there ever have been though the proportionate contribution of agriculture to national incomes has fallen significantly over the last 35 years (Hill, 1997). This has resulted in a piling up of marginally productive people in the agricultural sector. In Viet Nam, for example, it has been estimated that only 30-40 per cent of farmers can produce a surplus for sale (Hill, 1998).

Crystal ball gazing is an inherently risky operation. It is possible, for example that the current Asian financial crisis will deepen. If so the pace of change in marginal areas may slacken. Already, for example, it seems likely that the Electricity Generating Authority of Thailand will now not proceed with planned hydroelectric power stations on remote Laotian tributaries of the Mekong River. However, the crisis seems unlikely to do more than slow the pace of globalisation, structural change and thus their impact on marginal regions. Such a slowing must have its effects too, for tourism is diminishing, the lowland and urban labour markets are contracting and these processes will tend to exacerbate some of the trajectories to be outlined here.

First, though, it is important to set a future time frame for these trajectories. Half a lifetime seems reasonable if only for the reason that percipient observers of the region, were, 30 years ago, suggesting that provided political issues could be settled, it would take off into rapid capitalist development, that overall there would be significant real growth. Arguably, that growth has been partly shared by at least some of the region's marginal peoples. The remotest villages now use kerosene lamps and many have small generators to supply power for lighting, and the transistor radio is widespread even if some families do not own one.

Status Quo

The first trajectory to be considered is perhaps not a trajectory at all, or at best a weak one. It is the maintenance of the status quo. What is this? It can safely be asserted that with the exception of a few extremely remote, mainly forest dwelling groups, populations are growing at moderate, sometimes rapid rates, basically as a result of improved water-supply, hygiene and access to simple health care, though these vary from place to place. Permanent, annual sloping-land cultivation, shifting cultivation, supplemented by hunting and gathering, sometimes commercially-oriented, or by wet rice production which, though limited by a lack of suitable land, has clearly expanded, remain the basis of the local economies. Where there is access to wage labour, as in northern Thailand, or to as yet little-used, though not necessarily original forest, population growth does not result in either a serious shortening of forest fallows (below about 10-12 years) or a shift to permanent forms of agriculture. Temporary, usually seasonal out-migration – to work (until recently) on urban construction sites (Thailand) or in logging camps (Sabah and Sarawak) – provides employment for cash, allows the purchase of small 'luxury' items, as well as improved access to education and also permits much of the traditional way to survive. In some instances, though, the loss of labour has resulted in a reduction of areas cleared, reducing local food production, which is compensated by the importation of lowland rice, cooking oil, and tinned food. The community no longer sustains itself from the local environment, though the pressure on that environment may be somewhat reduced, except in respect of wild game now hunted with dogs and shotguns rather than by blowpipe and pitfall. Whether, in the longer run, temporary migration will be replaced by permanent migration to the lowlands and urban areas, followed by actual depopulation of the hinterland is not clear but the example of the Kelabit highlands of Sarawak suggests that remoter village sites may be abandoned in favour of those more accessible, not least to a small but growing eco-tourist trade.

The downward spiral of permanent sloping-land cultivation

A second trajectory concerns permanent or semi-permanent cultivation of annual crops on sloping land. Just how widespread this system may be is unclear but examples include dry rice growing in Java, the *padi gogo* system, the cultivation of manioc, either for subsistence, as in the Gunung Kidul region of Java or for commerce, as on the rolling uplands of northeast Thailand. In both these some degree of fallow is employed not least because yields fall to uneconomic levels after 3-6 years of cultivation and/or weed problems necessitate fallow. Not so in Guizhou (Hill, 1993a; 1993b) and parts of Yunnan where on a high proportion of the sloping land two crops are taken every year, rice in summer and rape-seed or wheat in winter, mainly for family subsistence. In Guizhou for instance, the crop intensity index on sloping land is an astonishing 189 meaning that on average all land is cultivated once a year and 89 per cent of it is cultivated twice a year. In every province in southern China land formed into sloping terraces is used for short-term crops, which in warmer areas include maize, tobacco, sugarcane and vegetables.

Given that conservation measures such as contour ploughing, terrace-edge bunds or erosion-control hedges scarcely exist, the question of sustainability must arise. Measured annual soil losses range mainly from about 1000 to 5000 t/km^2 with sometimes up to an order of magnitude higher on some deep-weathered soils such as those on basalt or andosols developed on volcanic ejecta. Just how far and how fast such eroded materials move is largely a matter for conjecture for much may be retained in or near river channels as point bars, bankside and channel-floor deposits mobilised only during major rainstorms and floods. During such episodes, however, the deposition downstream of half to one metre of material is not too uncommon at or near the usual break of gradient in the piedmont zone, with centimetres of material lower down. Since such torrential deposits may include significant quantities of sand, gravel, even boulders, the effects of the washout of accumulated sediments extend far beyond their place of origin in the hills and mountains. It is highly significant that in north-western parts of southern China lying within the Yangtze catchment are seeing vigorous afforestation aimed at reducing the rate at which the Three Gorges reservoir will accumulate sediment. However, since farmers whose land is being afforested still have to obtain food from it even the medium-term success of such action is problematic. The real issue is how to leave the land in production while also reducing the generation of sediment by surface wash and slope failure, the latter perhaps an impossible task.

The downward spiral of shifting cultivation

A third trajectory leads in the direction of the long-familiar downward spiral characteristic of growing population, limited land, over-exploitation of forest resources, reduced forest fallows, reduced crop yields leading to further reduction in fallow (Nye and Greenland, 1960). It has also long been thought that the abandonment of swiddens in the region comes about because of the depletion of nutrients after a year or two of cropping on soils that are mostly low in nutrients initially. However the matter has never been systematically put to the test in the region and it is at least as likely that regeneration of the vegetation, some kinds of which (*Imperata* grass is one) can flourish even at low nutrient levels, is so vigorous that an inordinate amount of labour would need to be expended on weeding in order that cultivation could continue (Hinton, 1978).

In a situation of growing population, limited out-migration and reducing fallows two scenarios ensue. In both cases the system comes close to collapse, but with differing responses. The first response is removal elsewhere, onto more or less undamaged land, often in forest reserves. This is the characteristic response of the highly mobile Hmong people of northern Viet Nam and the southwestern Chinese borderlands. Such groups may, of course, come into conflict with other groups occupying or using land to which, at least by tradition, they have prior right or, of course, with representatives of central authority such as forest guards, border police or the military.

The alternative response as fallows decrease and the system nears breaking point is to adopt some form of permanent agriculture. This does not necessarily imply biting the bullet of enhanced labour and/or fertiliser inputs to maintain crop yields. In the wetter, western parts of insular Southeast Asia – Sumatra, the Malay Peninsula, southern Thailand, a few small areas in Java, Borneo – the response has been to add *Hevea* rubber as a cash crop (Jensen, 1966). In other equatorial areas, notably in the Philippines, the tree-crop of choice is coconut. For continental Southeast Asia and southern China a move to perennial tree-crops is much more difficult. Coconuts yield poorly or not at all, being close to their northern climatic limits. Rubber will grow in southern China though not well, but depends upon prices set above those of the open market to be worth growing at all. In any case, rubber has rarely been adopted as a smallholder crop in that sub-region and I know of no case in which shifting cultivators have attempted to grow it. Both coffee and tea have been adopted by such cultivators. Coffee from the central highlands of Viet Nam, where it is grown on superb basalt-derived soils by non-Vietnamese peoples, is notably good. But none finds its way onto international markets and the local market, in a predominantly tea-drinking country, is relatively small. Where access to urban even export markets is possible, for example, near Chiang Mai, northern Thailand vegetables and cut flowers have been successfully added to the crop assemblages of people now wholly or substantially sedentary cultivators. Citrus, litchi and on parts of the high plateau of southwestern China, apples may be similarly added but the serious difficulty is that there is simply no perennial tree-crop (other than tea and coffee for which demand is limited) which is non-perishable, of small bulk and of sufficient value to bear the considerable costs of overland transportation, by human or animal porterage, in many cases.

One colonial-period answer to this dilemma was the still-surviving *taungya* system of northern and eastern Burma. This involved the raising and planting of teak (*Tectona grandis*) seedlings in swiddens under the general supervision of forestry officials. Teak has a fairly open crown allowing it to be intercropped with food crops quite successfully. From colonial Burma the system was transferred elsewhere, notably to parts of Java, where it also survives though in the recent past only with some degree of government subsidisation, especially for fertilisers (Jordan, Jiragorn Gajaseni and Watanabe, 1992). The major problems with this approach are, first, that teak has very specific environmental requirements, limiting its application to relatively small areas in upland Burma, northern Thailand and Java, and second that teak has a rotation period of 80 years, rather too long for most upland peoples whose control of land is based upon custom rather than legal writ.

A further answer to the dilemma of finding high-value, low-bulk crops, one which preserves, even enhances the livelihood of traditional shifting cultivators in continental Southeast Asia, and possibly now in the Chinese borderlands, is the (illicit) cultivation of opium. Small amounts are locally consumed either as opium or its derivative, heroin, but the overwhelming proportion, estimated to be about 90 per cent, is exported to meet foreign demand. Opium grows well in swiddens and easily bears the costs of transportation (Gutelman, 1974). Its main centre of production remains in the infamous Golden Triangle though the contribution to production by Thailand has recently fallen substantially as a result of suppression and to a limited degree because of crop substitution programs. Interestingly, another though lower-value illicit crop, cannabis, has not, it seems, been adopted on a significant scale by upland peoples, though it grows well. Presumably, the market is well supplied by lowland growers.

An upward spiral?

Is there, then, likely to be another trajectory, an upward spiral of improvement for marginal peoples in the region? Much depends upon what is meant by 'improvement'. For some politically dominant majority groups 'improvement' clearly means abandonment of what are seen to be 'nomadic' lifestyles (Hill, 1985), and ultimately assimilation to the majority. Marginal people are seen as 'backward' or 'inferior'. (Until the 1950s the Chinese names of minority groups in southwestern China generally bore within the language the radicle for 'dog'.) If their culture is to be preserved at all it is as a human zoo such as is to be found near Guiyang, the capital of China's Guizhou Province, where prostitution is to be found within the context of minority people's 'model villages' constructed for the entertainment of visitors. (Outside the region similar 'zoos' are to be found in Taiwan and Hokkaido, Japan.) Dance and the production of tourist geegaws alone are deemed worthy of preservation, with consequential reduction of the minority people to dependent status, not that this is by any means a new phenomenon (Reid, 1987).

An upward spiral of improvement would see the stabilisation of shifting cultivation at sustainable levels, i.e. fallows of a minimum 10-12 years, stabilisation of hunting and collection of forest products such that there is no long-term decline in numbers or biodiversity. It would see the development of tourism on the terms of and under the control of minority peoples themselves. Finally, such a scenario would probably see the continuance of a degree of dependence upon remittances from 'detached' members of communities as well as seasonal employment in towns and cities.

Does such a trajectory already exist in the region and could it develop where it does not? One small-scale example is that of Aeta minority people based near Subic Bay, Luzon. They have developed tourist-based activities centred upon Mount Pinatubo capitalising on continuing interest in the aftermath of the 1994 eruption but including a 'survival in the jungle' experience. Other examples are the well-established Mount Kinabalu tourist trek in Sabah, East Malaysia, and the more recent Mulu Caves tourist development in Sarawak, though in the latter case the luxury hotel component is controlled by an 'outside' interest, as is air access. Significantly though, levels of participation by indigenes has in these cases been founded upon a degree of literacy in English and a willingness to promote local interest, a seriously difficult matter in the face of the access to expertise and capital obtainable by majority-group and international tourist interests.

What form would agriculture take? An example is the substantial development of temperate vegetable production in the Cordillera Central north-east of Baguio, Luzon, where shifting cultivation has been suppressed both legally and by economic competition (Reyes-Baguiren, 1989). The area is just a day's journey away from the very large and comparatively wealthy Manila market. The shift to vegetables, however, has brought in its train serious environmental problems – local soil and water toxicity due to excess use of artificial fertilisers, herbicides and pesticides, soil erosion and sometimes also total economic dependence of farmers upon vegetable merchants. Similar but smaller developments include those near Kundasang, on the slopes of Mount Kinabalu, and at west of Chiang Mai. (Yet other temperate-crop developments such as around Bogor, Java, Cameron Highlands, Peninsular Malaysia, Tam Dao, northern Viet Nam, Da Lat, southern Viet Nam, are largely operated by lowlanders, not former minority-group shifting cultivators.) Other possibilities for the future might include orchid cultivation, also the development of local cotton for clothing for sale to tourists, already begun at a few places in northern Thailand, but they require access to markets, skills and capital largely beyond the present capacity of local people to mobilise them. Upland and other marginal peoples have been so long in dependent relationships with mainly lowland, majority peoples that autochthonous improvement is likely to be extremely limited though not impossible, if government policies include affirmative action while minimising dependence. But in most of the region that is probably too much to hope for.

References

Gordon, C.T., Gajaseni, Jiragorn and Watanabe, H. (eds.) (1992), *Taungya, forest plantations with agriculture in Southeast Asia*, CAB International, Wallingford.
Gutelman, M. (1974), L'économie politique du pavot à opium dans le Triangle d'Or, *Etudes rurales*, 53-56, pp.513-526.
Hill, R.D. (1985), Primitives to peasants? The 'sedentarization of the nomads' in Viet Nam, *Pacific viewpoint* 26(2), pp.448-459.
Hill, R.D. (1993a), Land, people and an equilibrium trap: Guizhou Province, P.R.C., *Pacific viewpoint* 34(3), pp.1-22.
Hill, R.D. (1993b), The unsustainability of agriculture in Asian tropical uplands under population pressure – some policy implications, *Proceedings of the ASAIHL Seminar on Sustainable Development, challenges for the Asia-Pacific region in the 21st century*, Griffith University, Nathan, Brisbane (publ. on disk).
Hill, R.D. (1997), An agricultural transition on the Pacific Rim: explorations towards a model, in R.T. Watters and T. G. McGee (eds.), *Asia-Pacific: new geographies of the Pacific Rim*, Hurst, London, pp.93-112.
Hill, R.D. (1998), Stasis and change in forty years of Southeast Asian agriculture, *Singapore journal of tropical geography* 19(1), pp.1-25.
Hill, R.D. and Peart, M.R. (in press). Land use, run-off, erosion and their control: a review for southern China, *Applied hydrology*.
Hinton, P. (1978), Declining production among sedentary swidden cultivators: The case of the Pwo Karen, in P. Kunstadter *et al.* (eds.), *Farmers in the forest*, East-West Center, Honolulu, pp.185-198.
Jensen, E.H. (1966), *Money for rice, the introduction of settled agriculture based on cash crops among the Iban of Sumatra*, Danish Board for Cooperation with Developing Countries, Copenhagen.
Nye, P.H. and Greenland, D.J. (1960), *The soil under shifting cultivation*, Commonwealth Agricultural Bureaux, Farnham Royal.
Reid, A. (1987), Low population growth and its causes in pre-colonial Southeast Asia, in N.G. Owen (ed.), *Death and disease in Southeast Asia, explorations in social, medical and demographic history*, Oxford University Press, Singapore, pp.33-47.
Reyes-Boguiren, R. (1988), *The history and political economy of the vegetable industry in Benguet*, Cordillera Studies Center, Baguio.

Acknowledgements

I am grateful to Professor W.J. Kyle for certain information contained in this paper and to the Conference organisers for the opportunity to prepare and present it.

4 Economic scenario of the reclamation and utilisation of marginal lands in India

ABHA LAKSHMI SINGH

Introduction

The fundamental problem, which the world faces today, is the rapidly increasing pressure of population on the resources of land. Today India is facing the same problem with the tremendous growth of population expected to swell to 1,000 million by the turn of the century. Large parts of our country are lying waste due to various reasons such as, salinity and alkalinity, soil erosion, water logging, weeds infestation etc. All these lands are characterised by some kind of deficiency and non availability of micronutrients (Singh,1985).

It is estimated that out of total geographical area of India, about 150 million hectares are seriously affected by erosion, wind or water, about 15 million hectares by salinity and water logging, about 20 million hectares by floods and another 20 million hectares could be added to this list due to the inadequacy of distribution system and lack of proper arrangements for draining excess water. This brings the area to be improved to a total 205 million hectares, 62.5% of the total geographical area (Government of India, 1986).

The marginal lands after reclamation could be utilised in various ways if a detailed knowledge about the conditions in these areas is available before hand. They could be brought under cultivation, forests could be grown on them, pastures could also be developed, settlements and communication lines could also be developed and so on. Hence, marginal lands after reclamation could be utilised in various ways. India is a vast country with a population of about 990 millions (1998) and this population is rapidly increasing. So the increasing demands of food and other products is a compelling reason for the further utilisation of these marginal lands to their maximum limits. This calls for micro level detailed study of areas, which have remained unproductive and unattended. Wherever attempts have been made to reclaim these marginal lands, critical examination must be made of the methods adopted for the development of the area, input-output balance sheet of the expenditure involved etc.

This paper on the 'Economic Scenario of Reclamation and Utilisation of Marginal Lands in India' has been divided into three parts. The first part deals with the concept of marginal lands; the second part deals with saline-alkali lands, its distribution in India, causes of lands becoming saline and alkali, its reclamation and further utilisation. While the third part deals with a case study sighted from Aligarh district where such types of marginal lands are being reclaimed and utilised. The cost of reclaiming such lands has been calculated.

Concept of marginal lands

The term marginal lands was used by the old writers to indicate the little used common lands usually on light soil which has failed to yield a return to the cultivators (Stamp, 1963). But now these open spaces have become of much value especially in the urban areas where there is scarcity of open areas. To the lay man this term has little meaning. He defines it as that land which is lying waste and serving no purpose. Thus, marginal lands may be defined as those lands, which are not available for cultivation or are left out of cultivation like fallows and cultivable waste. It embarrasses, first, lands not available for cultivation, barren and uncultivable waste, second, other uncultivated lands excluding fallows, cultivable waste, permanent pastures and land under miscellaneous trees, and finally, third, fallows, four, broadly marginal lands include all types of land thrown out of cultivation, providing little or nothing; a wild desolate region; a desert; a barren land, five, further, marginal lands could be broadly divided into two categories, six cultivable waste, seven uncultivable waste (Planning Commission, 1963).

Cultivable waste lands are those land which have the potential for further development of vegetative cover and it is lying unused due to different constraints of varying degrees like deep rooted grasses and weeds, unhealthy conditions, lack of drainage, lack of water supply, salinity and alkalinity, low fertility of soil and damage to soil by wild animals. The uncultivable wasteland can not be developed for vegetative cover and it includes barren rocky stony waste rock area, steep sloping area and show covered area. These marginal lands may be natural and also man made. The natural causes may be water erosion, wind erosion, leaching, scorching, water logging, change of river courses, tidal waves etc. The man made causes include deforestation, changing natural slope of drainage, over use of irrigation, chemical fertilisers, non rotation of crops, fallow lands, major and minor mining, urbanisation, changing land use like filling up ponds, water areas, depasturisation for commercial use, construction like roads, railways, new colonies embankments etc.

Saline-Alkali lands

According to a latest estimate 15 to 20 million hectares of land are engulfed with the problem of salinity and alkalinity. This is not a new phenomena but saline and alkali lands have known to occur in India from time immemorial. Most of these lands were used as grazing sites in the villages. It is recorded that till 1800 AD much of these lands were covered with some type of *Butea* forests that provided both forage and fire wood to the villagers Gradually with the pressure of population on land and increasing demand for fire wood, cultivation was started on these lands by cutting down of forests which resulted in denudation and ecological imbalances and construction of large scale embankments all around the roads, railways and canals. This led to the obstruction of natural drainage and this helped in the extension of areas under salinity and alkalinity.

With the view to improve agriculture and increase production, irrigation was introduced. This brought large-scale conversion of good lands into saline-alkali lands. Canal irrigation led to indiscriminate use of water, and helped in raising the ground water level. This led to the creation of swamps, water logging, accumulation of salt efflorescence and finally an increase in salinity and alkalinity. A growing concern for this problem started in 1886 in India and important investigations were made by soil scientists. The Government of India also realised the importance of systematic research on various aspects of salinity and alkalinity.

The salt affected lands are known by different names in different parts of the country. This basic reasons for this is that the conditions causing the formation of such lands differ widely from country to country and from region to region. Similarly its classification also varies. A widely accepted classification given by the U.S. Salinity Department (US salinity Lab, 1954) on the basis of chemical system is following, saline soils, alkali soils and saline-alkali soils.

Saline soils contain sufficient soluble salts to impair its productivity. If the solution extracted from saturated soil paste has ECe (Electrical

Conductivity) value of 4 or more millions per cm at 25°C, the ESP (exchangeable sodium percentage) is less that 15 and pH is usually less than 8.5, the soil is saline. The commonly found salts are sodium sulphate, magnesium sulphate, calcium sulphate, magnesium chloride, calcium chloride, sodium bicarbonates and nitrates.

Alkali soils may be defined in terms of productivity as influenced by exchangeable sodium. It may or may not contain soluble salts. The amount of ESP present in alkali soils is more than 15, the pH is between 8.5 and 10 and in some cases it exceeds 10 and the ECe is less than 4 million per cm at 25°C.

The saline-alkali soil is that soil which is suffering both from the problem of salinity and alkalinity. The ECe is more than 4 millions per cm at 25°C, the ESP is more than 1.5 and the pH is generally above 8.5.

In India, the salt affected lands are known by a variety of names but the most common name is *usar* which means barren lands. It is usually applied to all types of saline-alkali lands, mostly found in arid and semi-arid regions of the country. It is more pronounced in the states of Uttar Pradesh, Punjab, Haryana, Rajasthan, West Bengal, Orissa, Maharashtra, Karnataka and Madhya Pradesh. The four major tracts are; first, in the semi-arid Indo-Gangatic alluvial tract mainly in Punjab, Haryana, Uttar Pradesh and parts of Bihar; second, arid tracts of Rajasthan and Gujarat; third, arid and semi-arid tracts of the southern states mainly in the irrigated black soil region; four, the coastal alluvium.

A combination of factors which are mainly geological, climatic, hydrological in value is usually involved in the formation of these degraded soils. The intimate relationship of geology is often reflected in the nature of soluble salts in the weathering crust of rocks of the parent material from which the soil originated. The arid and semi-arid climates where precipitation is usually less than evaporation associated with certain elements of topography and ground water hydrology are often responsible for the accumulation or transport and deposits in other places of salts.

Weathering is the primary source of all kinds of soluble salts usually found in salts, contemporary weathering and soil forming process can be regarded as a major significance in salinisation of soils. The source of salts in many cases is more secondary deposits such as shales, sand stones, glacial and wind borne materials and unconsolidated alluvium of various geological ages. Saline soils are exclusively part of the arid zone landscape. Rainfall appears to have greater effect than annual temperature in delimiting the salinity of the soil soluble salts accumulates wherever evaporation exceeds precipitation, either alone or in combination with irrigation. Salt accumulation is also associated to low lands and to regions of high water table. At such places the surface run off is negligible and drainage water evaporate leaving the salts on the surface. The process of evaporation also brings up salts from the lower horizons, if the moisture regime is connected with the ground. The use of saline water for irrigation has accentuated the problem of bringing salts in extreme areas. Irrigation by canals has also been the principal cause of extension of this problem. It also helps in raising the water table.

Economics of reclamation of saline-alkali lands

Before discussing about the economics of reclamation we should first focus our attention as to how these lands are reclaimed since these lands differ in their characteristics, the methods to reclaim them also differ. The methods adopted must be based on the proper causative factors, which led to the development of such soils. Depending upon the local conditions, a combination of techniques is usually required to achieve a personal solution for the problem (Singh, 1978).

Generally establishing drainage and resorting to leaching operation with enough quantity of good quality irrigation water reclaims saline soils. Alkali soils in addition require a suitable amendment for ready source of soluble calcium to neutralise the alkali, replace the exchangeable sodium in the soil and flocculate the clay improving permeability. Such a source of calcium is found in a number of ways either by adding directly a soluble calcium chloride or using an acidifier (organic or inorganic) capable of dissolving the natural insoluble calcium. Thus, saline-alkali soil requires treatment with gypsum, organic matter, green manuring, flooding, flushing and leaching with the help of good quality and large qualities of water, the improvement of drainage, the soil management involving land levelling light tillage operations, judicious fertiliser management, water management and the establishment of vegetative cover. Broadly speaking the reclamation methods could be grouped as, physical and hydro-technical amelioration, biological amelioration, and chemical amelioration. The physical hydro-technical amelioration includes, mechanical treatments like deep ploughing, sub-soiling, sand filling, profile inversion, breaking the hard pan through the use of diesel tractors etc. Leaching and drainage constitute the hydro-technical part. Biological amelioration includes addition of organic matter, growing of crops, grasses, green manuring, addition of crop residues, farmyard manures, preesmud. Chemical amelioration are gypsum, lime stone, rock phosphate, acidifying material such as sulphur, iron sulphate, iron pyrites, aluminium sulphates.

In actual practice the reclamation is made more effective and speedy by combining various ameliorative measures. Management of such soils include mechanical and agronomic practices which help to keep down the harmful effects of salts and alkali during and after the reclamation period to ovoid re-salinisation of lands to its original status. Suitable management include adoption of special planting and irrigation techniques, afforestation and pasture development, cultural practices, land preparation and tillage methods.

In calculating the cost of reclamation of saline lands, saline-alkali lands and alkali lands, it will be seen that some of the processes are common to both types of soils. Operations like levelling and bunding, land preparation, transplantation, of paddy remains the same. In alkali soils gypsum is applied in larger quantities than in any other type while saline soils require leaching to a great extent. The application of fertiliser differs in quality and quantity. Cost of reclamation thus varies from field to field. This in view, and prevalent cost of labour and commodities, the cost of reclaiming these types of problematic marginal lands (Table 4.1) (Singh, 1978).

Table 4.1 Cost of reclaiming one acre of saline soils in Aligarh District, India

S.No.	Operation	Rate (Labour Materials)	Cost (in Rs*)
1.	Levelling and bunding	6 Labourers (Rs 40 per labourer)	240.00
2.	Land preparation	3 Labourers (Rs 40 per labourer)	120.00
3.	Draft employed	28 Hours (Rs 20 per hour)	560.00
4.	Leaching	Summer 6 Days (7 Hour per day) (Rs 20 per hour)	840.00
		Winter 4 Days (7 Hours per day) (Rs 20 per hour)	560.00
5.	Compost manure	Requirement 1000 kg	200.00
6.	Transplanting of seedling including cost of nursery		2,000.00
Total			4,520.00

* 1 US $ = Rs 42.50 (July 1999).

Marginal lands with saline encrustation (saline soils) have high soluble content so much so that during the dry months from February to June, white encrustation and white powder appears on the soil surface. These are best reclaimed with the provision of abundant good quality of water for leaching the salts and improving the drainage. Table 4.1 shows the cost of reclaiming one acre of saline soils in India comes to be Rs 4,520. Table 4.2 is showing the cost of reclaiming saline and alkali soils, which comes to be the Rs 9,627 for one acre. Table 4.3 is showing the cost of reclaiming one acre of alkali soils, which comes to be Rs 12,257.

Table 4.2   Cost of reclaiming one acre of saline-alkali soils in Aligarh District, India

S.No.	Operation	Rate (Labour Materials)	Cost (in Rs)*
1.	Levelling and bunding	6 Labourers (Rs 40 per labourer)	240.00
2.	Land preparation	3 Labourers (Rs 40 per labourer)	120.00
3.	Draft employed	28 Hours (Rs 20 per hour)	560.00
4.	Gypsum application	Requirement 3.5 tons (Rs 1200 per ton)	4200.00
	Labour employed	5 Labourers	250.00
	Average transportation cost		500.00
5.	Impounding of water	10 Hours (Rs 20 per hour)	200.00
6.	Fertilisers		
	(i) Urea	Requirement 50 kg (Rs 176.0 per kg)	176.00
	(ii) Shakti khad	Requirement 50 kg (Rs 458 per bag)	458.00
	(iii) M. of potash	Requirement 35 kg (Rs 205 per bag)	143.00
7.	Transplanting of seedlings including cost of nursery	2 Labourers (Rs 40 per labourer)	2,000.00
8.	Irrigation	35 Hours (Rs 40 per hour)	700.00
Total			9,627.00

* 1 US $ = Rs 42.50 (July 1999)

Table 4.3   Cost of reclaiming one acre of alkali soils in Aligarh District, India

S.No.	Operation	Rate (Labour Materials)	Cost (in Rs)*
1.	Levelling and bunding	15 Labourers (Rs 40 per labourer)	600.00
2.	Land preparation	5 Labourers (Rs 40 per labourer)	200.00
	Draft employed	56 Hours (Rs 20 per hour)	1120.00
3.	Gypsum application labour and transportation	Requirement 5 tons (Rs 1200 per ton)	6000.00
4.	Impounding of water		200.00
6.	Fertilisers		
	(iv) Urea	Requirement 50 kg (Rs 176.0 per 50 kg)	176.00
	(v) Shakti khad	Requirement 50 kg (Rs 458 per 50 kg)	458.00
	(vi) M. of potash	Requirement 30 kg (Rs 205 per bag)	123.00
7.	Irrigation		780.00
8.	Transportation of seedlings including nursery cost		2000.00
	Total		12,257.00

* US $ = Rs 42.50 (July 1999)

It is seen from the tables that large amount of money is involved in reclaiming these lands. This is beyond the reach of Indian farmers the Government should rescue these poor farmers.

Utilisation of reclaimed marginal lands

To assess the utilisation of reclaimed marginal lands field studies were conducted in villages where utilisation of such lands is being done (Singh and Hashmi, 1989). About 141.68 hectares of saline-alkali lands, spreading in the villages of Gursikran and Mahuakhera in Aligarh district was adopted for reclamation by Action For Food Production Project in 1983. This land is about 10.5 km away from the district headquarter.

The total area of Gursikran village is about 522 hectares out of which the net cultivated area is about 219 hectares (42%) and about 136 hectares (26%) of the land is affected by saline-alkali conditions which is mainly found in the western portion of the village. The total population of the village is 1,853 persons of which there are about 228 farmers, 217 agricultural labourers and 59 marginal agricultural labourers. The total area of Mahuakhera village is about 357 hectares of which the net cultivated area is about 206 hectares (58%) and about 117 hectares (33%) of land is affected by saline-alkali soils, which is mainly found in the eastern, western and southern portion of the village. The total population of the village is about 840 persons of which there are about 76 cultivators, 74 agricultural labourers and 43 marginal agricultural labourers and other workers

Earlier no crop was grown on this *usar* patch, but by 1983, this patch was adopted for reclamation by Action For Food Production Project (AFPRO) workers and they were reclaiming it by applying 15 tonnes per hectare of gypsum. In 1983-84 about 17 hectares of land was reclaimed and *paddy* was grown in *kharif* season (rainy season crop). This was followed by wheat, *dhaincha (sesbaina aculeata* – a green manuring crop). Of all these crops wheat was found to be economical and readily adaptable to these soils. During the first year of reclamation, the average yield of paddy was about 31 quintals and that of wheat 21 quintals per hectare. During the second year of reclamation (1984-85) about 25 hectares of additional land was reclaimed. About 35 quintals of *paddy* and 25 quintals of wheat per hectare were obtained. In the third year of reclamation (1985-86) about 37 hectares of additional land was reclaimed. The average yield of paddy and wheat was about 41 quintals and 29 quintals per hectare. In some areas barley, oat and *berseem (Triflouim alexandrinum* – a green fodder crop) was grown on experimental basis.

During the *kharif* season of 1986-87, about 50 hectares of land was reclaimed and out of which 35 hectares was under paddy, 10 hectares was under *jowar* (big millets), one hectare was under *dhaincha* and sugar cane. In the *rabi* season oat, wheat, mustard and *berseem* was grown. Oat occupied 23 hectares of land while wheat, *berseem* and mustard covered about 15 hectares, 10 hectares and 2 hectares of land. Apart from this about 4 hectares of land was under *Karnal (Diplachne fusca (Linn) P. Beaur), Rhodes (sporobolus sp. chloris gayana) and Para grass (Brachiaria mutica)* which also helped in the process of reclamation. The benefits accrued were the following (Table 4.4).

Table 4.4 Utilisation and benefits of reclamation in Gursikran and Mahuakhera villages of Aligarh District

Total area to be reclaimed equals 141.638 hectares

Increase in cultivated area

Years	Land Reclaimed (hectares)	Net Cultivated Area in both the villages (hectares)	Percentage to the total area	Percent of increase in Net Cultivated Area
1986-87	50.000	425	60	6.0
	141.638	425	60	16.1

Additional grain production

Years	Land Reclaimed (hectares)	Yield (q/ha) Paddy	Yield (q/ha) Wheat	Production (quintals) Paddy	Production (quintals) Wheat
1983-84	17.000	31.00	21.00	527.00	351
1984-85	25.000	35.12	24.79	878.00	619.75
1985-86	37.000	41.32	28.92	1,528.84	1,070.04
	141.638	35.00	25.00	4,957.00	3,541

Employment benefits

Years	Land Reclaimed (hectares)	Human labour days (per/ha)	Total human labour requirement
1983-84	17.000	160	2,720
1984-85	25.000	150	3,750
1085-86	37.000	150	5,550
	141.638	150	21,245

Additional gross income

Years	Land Reclaimed (hectares)	Average yield (q/hectare) Paddy	Average yield (q/hectare) Wheat	Total production (quintals) Paddy	Total production (quintals) Wheat	Gross income (Rs) Paddy	Gross income (Rs) Wheat	Total income (Rs)
1985-86	37.000	35	25	1,295	925	186,480	149,850	336,330
	141.638	35	25	4,957	3,541	713,808	573,642	1,287,450

Additional net income

Years	Land Reclaimed (hectares)	Average net income (Rs) (per/hectare) Paddy	Average net income (Rs) (per/hectare) Wheat	Net income (Rs) Paddy	Net income (Rs) Wheat	Total income (Rs)
1985-86	37.000	1,000	250	37,000	9,250	46,250.0
	141.638	1,000	250	141,638	35,410	177,047.5

Table 4.4 continues

Additional input demand

Years	Land Reclaimed (hectare)	Demand for improved seeds		Demand for nitrogen fertilisers		Total
		Wheat (40 kg/ha)	Paddy (100 kg/ha)	Paddy (130 kg/ha)	Wheat (110 kg/ha)	
1985-86	37.000	1,480	3,700	4,810	4,070	8,880
	141.638	5,665.52	14,163.8	18,413	15,580	33,993

Benefits of dairy programme
a) Benefits from grass cultivation

Years	Total cost (Rs/ha)	Production (q/ha)	Returns (Rs)	Net profit (Rs)
1983-84	3,035	180	1,800	
1984-85	1,000	180	1,800	
1985-86	1,000	180	1,800	
Total	5,035	540	5,400	365

b) Benefits from milk production

Season	Production	Cost (Rs 4 per litre)
Winter	120 to 130	480 to 620
Summer	70 to 80	280 to 320

c) Employment benefits
about 27 labourers get job

10 labourers	24 hours	paid Rs 19 to Rs 21 per day
5 labourers	12 hours	paid Rs 12 per day
12 labourers	24 hours	paid Rs 19 to Rs 20 per day

* 1 US$ = Rs 42.50, July 1999

Source: Action for Food Production Project, Aligarh (1986), Compiled from field observations.

Increase in production

During the first year of reclamation, the yield of paddy and wheat were 31 and 21 quintals per hectare, second year the yield was 35.12 and 24.79 quintals per hectare and in the third year it was 41.32 and 28.92 quintals per hectare. There was increase in yields and this was due to suitable reclamation and farm management practices. During the first year, the production from 17 hectares of reclaimed land was about 527 quintals of paddy and 351 quintals of wheat. In the second year, the production of paddy was about 878 and that of wheat 619.75 quintals from the 25 hectares of reclaimed land. During the third year, the production of paddy and wheat from 37 hectares of reclaimed land was 1,528.84 and 1,070.04 quintals. This additional grain production has led to increase in per capita availability of food grains in the villages of Gursikran and Mahuakhera and this has further enhanced the economy of the two villages.

Employment benefits

Land reclamation has generated more employment to the rural labour force, bullock labour, tractor, tubewells and threshers These services are being used to reclaim land and subsequent crop production. Labour is required for land preparation, bunding, levelling, and application of amendments and for different operations of paddy and wheat cultivation. During the first year, land preparation and amendment application required 35 human labour days, cultivation of paddy required 85 human labour days and wheat crop required 40 human labour days per hectare. So during the initial year the entire tillage operations for paddy crop were covered under land reclamation thus reducing the labour requirements. But from the second year, the labour requirements for paddy increased to 95 days and that of wheat to 55 days per hectare. Thus during the first year the annual labour requirement was about 160 human labour days and from the second year it was about 150 human labour days per hectare. Since 37 hectares of land has been reclaimed, the annual labour requirement was about 5,550 human labour days.

Benefits of additional gross income

Income exceeds investment from the third year. The prevailing Government prices (1986) for paddy and wheat was about Rs 144 and Rs 162 per quintal ($1 U.S. = Rs 12.95, in 1986) of paddy worth Rs 186,480 and 925 quintals of wheat worth Rs 149,840 per annum. Thus, there was about Rs 336,330 per annum of additional gross income.

Benefits of additional net income

There was an additional net income of Rs 5,000 per hectare from paddy and wheat cultivation. Thus the total annual net income was about Rs 185,000 after three years of reclamation from 37 hectares of reclaimed land. This land increased the per capita income and has helped in the upliftment of the economic conditions of these villagers. The farming standards have also improved in the surrounding villages.

Benefits of increased use of inputs

The newly reclaimed land has created more demand for high yielding varieties, bullock, tractors, tubewells, fertilisers, plant protection measures etc. About 37 hectares of land has been reclaimed for crop production and this land needs about 40 kg per hectare of improved seeds of paddy and 100 kg per hectare of improved seeds of wheat per annum. Nitrogen fertilisers at a rate of 130 kg per hectare for paddy and 110 kg per hectare for wheat and there is an additional demand of 25 kg per hectare of zinc. Eleven tubewells have been installed for assured water supply and two tractors and two pairs of bullock are available for tillage operations. There is also demand for some other implements like threshers, spade, sickles and *kudalis* (used for digging) for the purpose of cultivation.

Benefits of dairy programme

About 4 hectares of land is being used for grass cultivation – *para, karnal and rhodes*. The *karnal* grass covers about 3 hectares while *para* and *rhodes* grasses each has half hectare of land under them. For the utilisation of these grasses, a dairy programme has been started and this has good scope because of two large milk consumers – the Central Dairy Farm (located in Cherat village) and Glaxo (located in Manzooorgarhi village) are present. There are about 53 Haryanvi cows, 21 Haryanvi female calves, 52 cross breed cows, one pair of bullock, two male buffaloes and seven male Haryanvi calves. The average milk production is about 100 litres per day, which varies from season to season. In winter the milk production is about 120 to 130 litres per day while in summers it varies from 70 to 80 litres per day. The milk is sold to the consumers at a rate of Rs 4 per litre and Rs 3.50 per litre to the workers of AFPRO project. The dairy programme also provides job opportunities to about 15 landless labourers Out of which 10 are employed for 24 hours for milking, feeding and taking care of cattle. These workers are paid Rs 19 to Rs 21 per day. The remaining five labourers are employed for daytime for cutting of grasses and jowar and for supplying it to the cattle. They are paid Rs 12 per day. Besides these, 12 more workers are employed at the project farm for different odd works and they are paid Rs 19 to Rs 20 per day. Thus about 27 landless labourers have got a permanent job and they are in addition to those who are employed for land reclamation and crop production purposes.

Conclusions

This study indicates that land reclamation is helpful in improving the socio-economic conditions of rural poor by providing more cultivated land, more pasture land, more employment opportunities, more income to rural people and finally it will not only solve the regional problems of rural life but will also help in solving national problems. The increased production will increase the per capita availability of food grains, increase the supply to transport agencies and processing units, the increased market arrivals will add to Government revenue in the form of market fee and this will create more demand for consumer goods and services. The increased demand for tubewells, fertilisers and implements will create new workshops in rural areas for repairing and supply of spare parts, it may encourage to open fertiliser, insecticides and seed stores and opening of these depots will further create new employment opportunities to rural skilled labour. Thus land reclamation will not only bring economic stabilisation but will also help in increasing the size of landholdings, increase employment opportunities and there will also be an increase of wage rate.

A large number of households in the villages do not own any land and they are landless labourers. They depend on seasonal work and are among the poorest in the rural community. They are also being benefited. The land during and after reclamation needs labour and this has increased employment opportunities for landless labourers in their native villages. Again because demand for labour becomes sharper than the supply, the wage rate has also increased (from Rs 10 per day in 1983 to Rs 15 per day in 1986). Now at many places the Government is distributing these reclaimed lands amongst the landless labourers.

Actual results obtained in the villages of Gursikran and Mahuakhera (see Table 4.4) of Aligarh district has shown that land reclamation has helped in solving food problems and bringing prosperity to the rural masses. Future projections for the whole country could be made (see Table 4.2) on the basis of these actual results. It is estimated that in India out of the total Geographical area of 328 million hectares about 7 to 12 million hectares (about 2.5% to 4%) of land are affected by salt or alkali (Jaggi,

1985). These soils occur in the semi-arid regions of the country and the development of such soils are more pronounced in the states of Uttar Pradesh, Punjab, Haryana, Rajasthan, Gujarat, West Bengal, Maharashtra, Orissa, Kamatka and Madhya Pradesh. The problem is particularly acute in and semi and tracts of Indo-Gangetic alluvial plains, which contains about 40% of the total salt affected area of the country.

It is quite difficult to foresee what grains and in what quantity will be grown if all the saline-alkali lands of the country are brought under cultivation. Field observations have revealed that the prevalent practise is to grow paddy and wheat, so taking the average yields for paddy to be 3.5 tonnes per hectare and that for wheat 2 tonnes per hectare, the additional production is estimated to be 72 million tonnes per annum. This will help in improving the economy of the individual farmers and will raise the standard of living. In addition, the newly reclaimed lands will generate additional employment, demand for improved seeds, fertilisers, bullocks, tractors, tubewell, etc. In India the problem of unemployment is more acute in rural areas than in urban areas. In rural areas there is the problem of landless labourers and marginal farmers. They have been forced to part with their lands because of new burdens and old debts. Today more than one-fifth of all the rural households owns no lands at all. However, the real solution would lie in generating employment. In any land reclamation programme labour force is required at all stages. On the basis of actual results obtained in Aligarh district, the annual labour requirement will be about 150 human labour days per hectare, there Will be an additional demand for 1.8 million human labour days per annum. Extrapolation from the actual figure shows that there will be an additional demand of 480,000 tonnes per hectare of paddy seeds and 1,200,000 tonnes per hectare of wheat seeds. The nitrogen fertiliser requirement for both the crops will be 2,980,000 tonnes per hectare. These demands will create new avenues for workshops and stores in rural areas. Again opening of such depots will create new employment avenues to rural skilled labour force. From these reclaimed lands the country will be getting an additional income from the production of crops. The prevailing prices for paddy is Rs 150 per quintals and that for wheat Rs 167 per quintals (1987 prices). The income will flow only after three years of reclamation. The actual net-income accrued in the villages of Aligarh district was about Rs 1,000 per hectare from paddy and Rs 250 per hectare from wheat, on the basis of this, it is projected that in India, the net income from paddy and wheat crops will be Rs 12,000 millions and Rs 3,000 millions per annum.

References

Government of India, Ministry of Information and Broadcasting (1986), *India, A Reference Manual*, The research and Reference Division, New Delhi.

Jaggi, T.N (1985), 'Iron Pyrites in Alkali Soil Reclamation', in *The Sunday Pioneer*, Lucknow, September 15.

Planning Commission, Committee on Natural Resources (1963), *Study on wastelands including saline and alkali and waterlogged lands and their reclamation*, New Delhi.

Singh, A.L. (1978), *Economic and Geography of Agriculture Land Reclamation*, B.R. Publishers, New Delhi.

Singh, A. L. (1985), *The Problem of Wastelands in India*, B.R. Publication, New Delhi.

Singh, A.L. and Hashmi, S.N.I (1989), 'Land Reclamation and Rural Development in Aligarh District, Some Projections', *Asian Profile, Caneda*, Volume 17, No. 5.

Stamp, L. D. (1963), *The Lands of Britain, Its use and Misuse*, London.

US Salinity Lab (1954), *Diagnosis and Improvement of salure and Alkali soil*, Handbook no. 60, Department of Agriculture, Washington.

5 The problem of decision areas within a fragile ecosystem: Uco Valley, Mendoza, Argentina

GLADYS MOLINA DE BUONO

Introduction

An actual academic concern is to find out how local aspects are influenced by global aspects, especially because nowadays appear new agents which modify the territory. It can be observed a territorial fragmentation due to multiple influences that diversify and produce overlapping decisions in the same place.

The purpose of this research is to show the difficulties found by management in asserting sustainable development, when institutions dedicated to aspects such as service supply, natural resources, production and social needs overlap in the same area. These institutions, which work independently from one another, and at the same time, increase in number and variety, usually promote contradictions, even archaic measures. All this results in great complexity for environmental planning and an overload of the duties and bureaucratic burdens for the inhabiting.

The study area contains three administrative units in Mendoza, Argentina, known as the Uco Valley. Public and private institutions are analysed and classified, and it is provided a set of maps for the jurisdictions and decision areas. This contribution allows us to speculate about the opportunities and risks faced by local communities, given a visible opposition between decentralised politics and bureaucratic attitude, between the organisation of solidarity nets and individual competitive decisions.

The Uco Valley

The Uco Valley is located in the central-west area of the province of Mendoza. It comprises three administrative units: Tupungato, Tunuyán and

San Carlos. The mountain in the west, which is part of the Andes range, features great unevenness, and heights between 4,000 and 6,000 metres. The climate is dry cold and snowy cold, with glaciers and little vegetation. Although it is a water reservation area, and it has great landscape value, it is marginal for economic activities and human occupation, due to its little accessibility. This natural area covers 50% of the surface.

In the east and north-east, the 'huayquerías' and hills surround the valley. These physical units have poor quality soil, lack of water, high incidence of linear erosion – wads – and a moderate continental semiarid climate. Although the most important economic activity is the oil exploitation and it occupies 30% of the surface it is considered rather passive for the economic and social life of the place.

Finally, the organised area covers only 20% of the surface, but it concentrates the human settlements and the economic activity. Topographically, it is enclosed and it comprises two parallel strips of land, the 'glacis' and the clear land. The 'glacis' is a piedmont with moderate slopes, run through by the Tunuyán river and its tributaries, infiltration area and aquifer reservoir, moderate cold continental semiarid climate. Its use capacity is restrained since some of its limitations, such as slope, weakness of the soil and the irregular distribution of rainfall are factors which generate floods, hydroid erosion, 'cárcavas', etc.

The clear land has better soil, flood material deposits and fluvial mud it is run by a fluvial fan constituted by the Tunuyán River and 18 tributaries of different origin. The climate is semiarid continental moderate. This area presents general weaknesses such as frost and hail. There are same places with specific problems such as floods, not very deep aquifer, and surfaces cuttered by deep and barracked rivulet. Its fragility is determined by the weaknesses already mentioned and by its dependence on the 'glacis', due to orientation of the rivers, the slopes and the soil.

These two strips of land, which are present in the three administrative areas, are the most dynamic from the human point of view. An expansion of the productive area to the west, to the 'glacis', is clearly observable, favoured by the technological changes that add innovations too quickly. Another process is the building of houses, which has produced an accelerated and disordered growth of cities, minor centres and rural neighbourhoods. The general impact is a backward movement of the traditional agrarian and ecologically worthy areas. At the same time, is also observable a residential advancement with empty sectors in the urban peripheries. This urban discontinuity only favours the land speculation and lessens environmental quality in general.

The institutions

The changes in public management, especially the decentralisation of administration, are producing a need of participation of the civil society. In turn, this fosters the organisation of new institutions, new Non Governmental Organisations (NGO) and the overall intervention of the community. But the lack of a joint territorial plan for the three administrative units, the overlapping of old and new institutions and the diversity of goals with which they work, are generating competition more than a harmonic development (see Figure 5.1).

Multiple influences at local scale

Some previous thoughts help as theoretical framework to order the different institutions, which share the same territory. The chart tries to represent the different influences each place receives simultaneously from the globalized world. To refer to these, Milton Santos talks about modernity and regulatory vectors, which create verticalities and horizontalities. That is to say, hierarchical relationships that connect the hegemony with the local and the everyday relationships created by territorial similarities and differences. However, in order to simplify this explanation and comparison with the role of the institutions in the Uco Valley, these factors will be mentioned as vertical and horizontal relationships.

The vertical relationships are established between different levels from the outside of a place. They may be *an abstract impersonal intervention*, with a tendency to generate a loss of identity since it levels up and universalises. It is the technology, the information and the urban culture. They may be too, *a concrete intervention* coming from the authorities or hierarchical institutions, exercised over perfectly delimited territories and jurisdictions. These are organisms either of the national or provincial government and therefore, the relationships are vertical and well adjusted to the corresponding legal framework. *Less frequent relation,* and in reverse order, is established as a vindicated reaction from the local community to a superior authority, ethnic movements, labour demands, and search of identity.

The horizontal relationships are agreed between territories and communities that share something in common. They can *be solidarity net among areas or discontinuous groups*, linked together by similar problems in different parts of the world. Intersection of the global and the local, non-

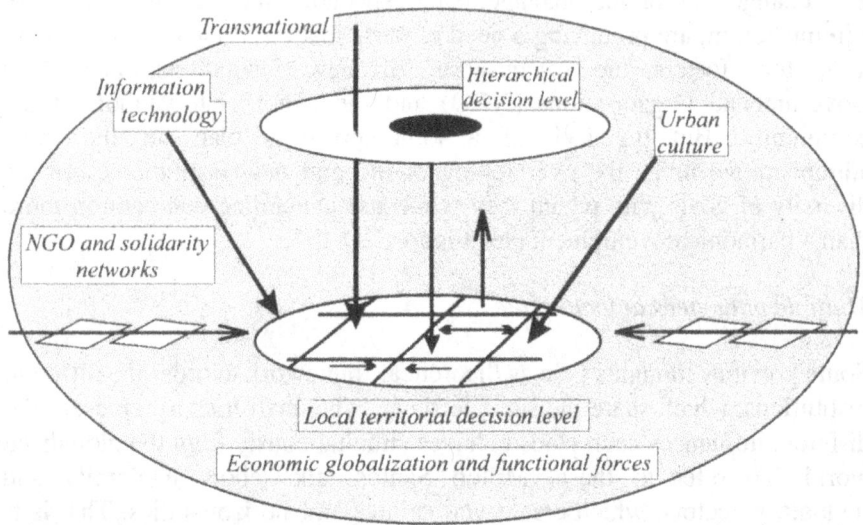

Vertical relationships

↓ Abstract imposition of global on local; impersonal relationships, identity loss

↓ Concrete imposition of national and provincial governments; political administrative and legal relationships

↑ Recovery reactions; local movements, ethnic groups, new status, search for indentity

Horizontal relationships

→ Solidarity network, intersection of global and local: groups with similar problems

←→ Local/regional territorial competition: the need to obtain benefits

→← Local/regional territorial solidarity: the need to join efforts through neighbour relationships

Figure 5.1   Territorial fragmentation: multiple influences at local level

governmental organisations, municipal networks sanitary humanitarian organisations, environmentalists, etc. Another horizontal relationships can be *a competition or confrontation* between local territorial interests, a need to attract benefits, an opportunity that is offered by decentralisation policies. Legal framework for a co-participation, royalties for natural resources exploitation, industrial promotion. Finally, *a territorial solidarity* generated by the need to join efforts through neighbouring relationships. Neighbourhood unions, productive co-operatives, patrimony defence organisations, commerce chambers, etc.

Territorial fragmentation and the institutions of the Uco Valley

There are a number of institutions that work in this place and their jurisdictions are very different, since they vary from a small neighbourhood to complete district or even the whole Uco Valley. They may be local or they may be a provincial delegation. According to the nature of their goals, they may be considered as control and government, service supply, consultation, information and support, or defence and solidarity. This last typology is more appropriate to develop the goal of this research. The list of institutions included (Table 5.1) summarises the data on territorial delimitation, location of the decision centre and gives some examples of the specific activities of each of them.

Control and government institutions

These entities represent the State, and their jurisdictions have limits, which are very rigid and subject to the legal framework. In this group are included the Town Councils, the Irrigation Department, the Provincial Road System Department, the Police Department and the National Frontier Guards. All these institutions are traditional but try to update their policies to adjust themselves to the changes, and to direct the processes of decentralisation and new management actions. Their activities are greatly affecting other institutions because they have the power of granting or banning, are typical of a 'megastate', which is the owner of society.

Table 5.1   Institutions and territorial influence of the various departments of the Uco Valley

Type of institution	Number of institutions	Location of decision centre	Territorial influence
Control and government	6 institutions with different limits	Local: 2 Extra local: 4	Rigid limitations or jurisdictions Functions: administration, control, permits, taxes, surveillance
Services supply	4 institutions with totally different limits	Local: 0 Extra local: 4	Clear limits but varied according to the spreading of the supply and its users Functions: distribution and commercialisation
Consultation and support	5 institutions with similar limits	Local: 0 Extra local: 5	Unfocused limits Functions: help in case of climate accidents; spread innovations
Solidarity and defence	86 institutions with non co-incidental limits	Local: 86 Extra local: 0	Clear limits for each community Functions: solve common interests; a joint to petition the authorities

Service supply institutions

The service supply institutions are the ones in charge of the distribution and administration of water, light, education and health through an adequate infrastructure, which increases the value of the places where they are spreading. They represent the 'welfare state' and comprise the Water Company, the Electricity Company, the General Education Board, the Regional Hospital, and health centres. All these institutions are being modernised by the incorporation of technology and privatisation processes. Therefore, their transformations also modify their relations with the society and the territory, since its areas of influence change in relation to the increase in the number of users of drinking water, sewage, power, gas, primary schools, etc.

Consultation, information and support institutions

They are fairly new institutions created during the last five years, with the aim of offering help to the producing sectors, preserving natural resources, inducing modernisation and favouring market connections. Although the goals are complementary, they depend on different ministries, that is the reason why they overlap their action in cultivated areas.

Solidarity institutions

They are free social organisations, which have existed for a long time, but are induced and reconsidered now. Community organisations, production chambers, western productive corridor association, other associations and co-operatives. Not all of them have clear spatial references, its areas are usually very flexible, even focused on other activities, but their action are multiplying. The current trend is to multiply and increase their protagonism. However, as they defend very small communities – a neighbourhood, a group of businessmen – they usually compete against one another looking for a chance to favour their micro territory rather than the whole Uco Valley.

It can be observed that many are branches of provincial institutions, therefore, they transfer instructions of different kinds from extra local authorities. The local institutions, most of them of the solidarity type, are too local, compete against each other, and relate with the authorities. There are not any local institutions that join the three departments; this increases uncertainty in the fragmentation of territorial interests (Figure 5.2).

From the territorial point of view, an overlapping of heterogenic control areas can be seen according to size: three districts, four police areas, one irrigation area, and one area pertaining to roads. The supply areas are complementary but all of them are within agrarian and urban surfaces. The promotion areas are the same that [coincide with) the agrarian surface. The solidarity areas, on the other hand, really produce a territorial fragmentation because they compete between towns, districts, communities, etc. The accumulation of local decision offices gives prestige to the town of Tunuyán which has 14 sites of different institutions, then follows La Consulta with five, Tupungato and Eugenio Bustos with four each.

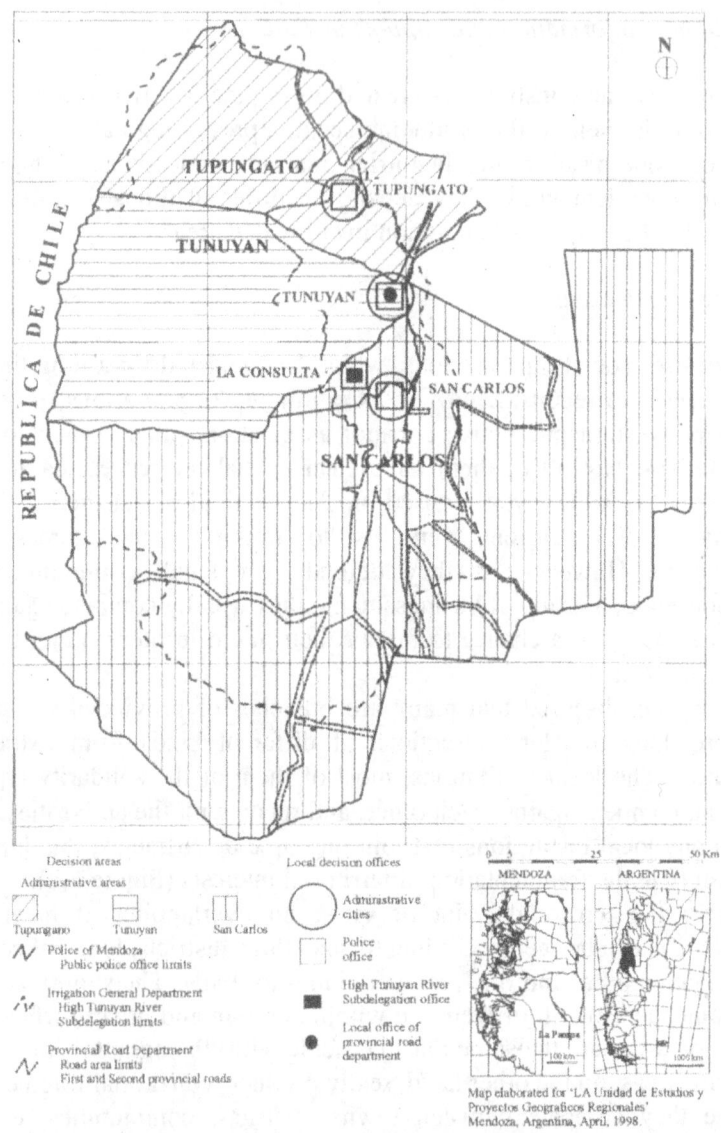

Figure 5.2 Control and government institutions, Uco Valley, Mendoza, Argentina

Source: Molina de Buono, G., Aportes para una Georgrafia Social, CONICET 1998.

Conclusions

The study of this place has allowed us to observe transformations in the decision areas, areas where new agents arise, overlapping their influence and modifying the territory. The maps illustrate a real territorial fragmentation that can be confirmed by consultation to the local communities. Among the positive aspects we can point out the social learning that improves civil responsibility, either by spontaneous organisation or by inducing participation practices from the government institutions. However, this constitutes an ecological risk since it increases the number of actors, introducing greater complexity to environmental plans.

Among the negative aspects, the most important is the absence of a territorial plan. While the institutions organise themselves, the spatial processes evolve with a high aleinatory degree. There is still lack of complementation concerning the inter-institutional relationships, which are bureaucratic rather than co-ordinating. Therefore, information is distributed in different offices, which difficult the studies and diagnoses. To conclude, if the fragmenting influence of globalisation cannot be prevented, a minimum of local conditions to preserve their own resources and values of the local community has to be imposed.

References

Barrio de Villanueva, P. (1996), 'Descentralización en Mendoza', in M. Furlani de Civit and M. Gutierrez de Manchón (eds.), *Mendoza: una geografía en transformación*, pp.233-257, Ex-Libris, Mendoza.

Gierhake, K. (1997), 'Presencia institucional y responsabilidades administrativas para la planificación ambiental: requisitos para el establecimiento de un sistema de gestión ambiental en Guatemala', *Revista Interamericana de Planificación*, vol. XXIX, n° 113, pp. 79-100.

Hirsch, J. (1997), 'Qué es la globalización', *Cuadernos del Sur* n° 24, edit. *Tierra del Fuego*, Buenos Aires, pp.9-20.

Medellin Torres, P. (1992), 'Reestructuración del estado y desarrollo regional: contrainsurgencia, democracia y disciplina social', *Revista Interamericana de Planificación*, n° 99-100, pp.87-101.

Molina de Buono, G. (1996), 'Las uniones vecinales en San Martín, Junín y Rivadavia. Aportes para evaluar su participación en el desarrollo local', in M. Furlani de Civit and M. Gutierrez de Manchón (eds.), *Mendoza: una geografía en transformación*, Ex-Libris, Mendoza, pp.259-280.

Molina de Buono, G. (1998), *Aportes para una Georgrafia Social*, CONICET.

Mosovich Pont-Lezica, D. (1997), 'Gestión del agua y organización de intereses en zona semiárida', in J.L. Luzón and T. Linch (eds.), *Regadío y Desarrollo en las regiones semiáridas americanas*, Universitat de Barcelona, Tarragona, pp.121-129.

Rodriguez Gutierrez, F. (1996), 'El desarrollo local, una aplicación geográfica. Exploración teórica e indagación sobre su práctica', *Ería, Revista cuatrimestral de Geografía* nº 39-40, pp.57-73.

Santos, M. (1993), 'Los espacios de la globalizacion', *Annales de Geografía de la Universidad Complutense*, nº 13, pp.69-77.

6 Agro-industrial enterprises and environmental fragility

MARIA JOSEFINA GUTIÉRREZ DE MANCHÓN AND
MARIA ESTELA FURLANI DE CIVIT

Introduction

This article presents results from an on-going research that concentrates on the issues of marginality as seen on society and territory transformations due to processes connected with economic opening, accelerated technical advances, internal decentralisation policies, and the consolidation of country blocks. Although we dealt with problems at the national level, the stress was placed on the regional and local scale. Thus Mendoza was studied as to the tendencies of foreign trade first, and then in what concerns the motivations and attitudes of agro-industrial enterprises, the necessary agents of an activity change. We also looked into the lack of complementation between external dynamics and territorial values. These values led us to the endogenous development, an attitude that favours local growth of culture and environment.

As a continuation of the previous subject we have selected a problem, both local and of present significance. In a reduced area, marginal by ecological risks, we are concerned with the difficulty of foreseeing the outcome of a spatial process of agrarian occupation, strong and accelerated, produced by the globalisation economic mechanism. We followed an intensive method, with fieldwork and interviews in depth.

The agrarian context

In Argentina extra-Pampean agricultural areas are characterised by their small size, discontinuity, need of total or partial irrigation in most cases, dependency on the internal market, and the structural problems of the 'minifundio'. In the province of Mendoza, two oases concentrate activity, the Northern and the Southern oases, activity that deals mainly with agro-industrial crops, vineyards, fruit orchards, and garden produce. The superior Tunuyán River basin, one of the sectors of the Northern Oases that stretches over three counties- Tunuyán, Tupungato, and San Carlos -is a hydrological reserve by the amount and quality of its superficial and underground waters. (Figure 6.1).

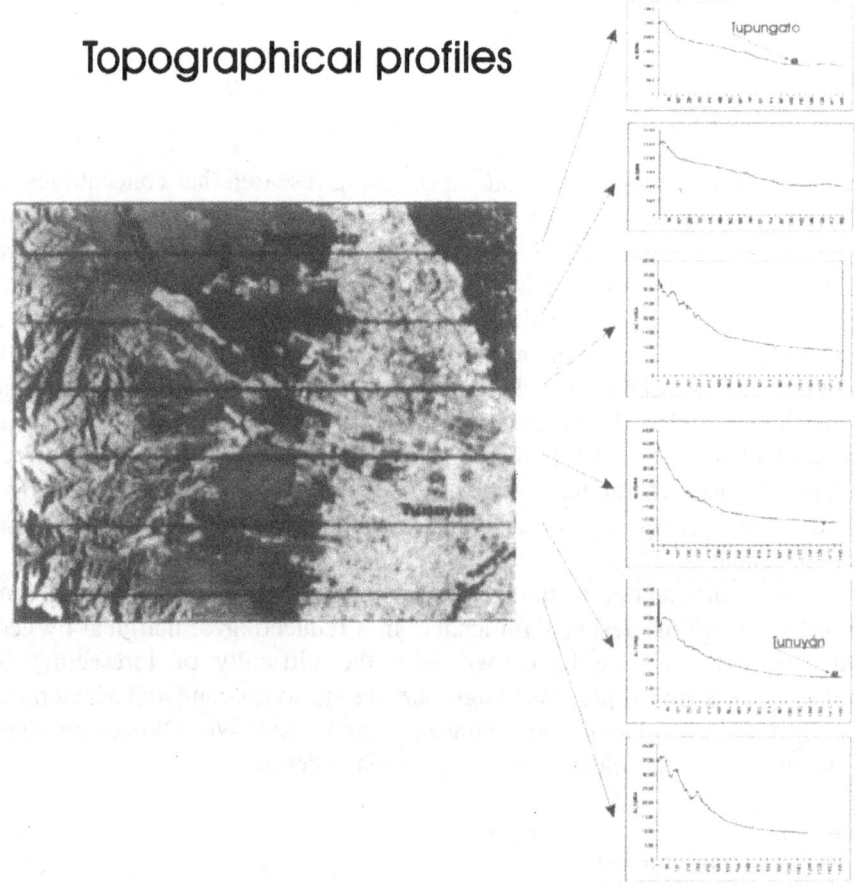

Figure 6.1   The study region: topographical profiles

The basin-organised space, known as Valle de Uco, composed of the piedmont (1,200 km^2), and the plain (1,270 km^2) is scarcely populated with areas apt for agrarian development and cattle raising. This basin has at present an agricultural identity characterised by diversification: fruit trees – apple, pear, cherry, walnut, peach and plum – vineyards and orchard produce – potato, garlic, carrot, tomato, onion, etc., – forestry and aromatic plants – oregano, mint, etc.

The understanding of the piedmont settlement process cannot be separated from the immediate agro-industrial situation. As was already stated agriculture under irrigation prospers in the Tupungato and Tunuyán plain. The data synthesise crop diversification, specialisation, and recent evolution (Tables 6.1, 6.2, and 6.3).

Table 6.1   Vineyards, in ha, 1971, 1988 and 1996

Regions	1971 [1]	1988 [2]	1996
World	9,870,000	8,727,000 [3]	7,768,000 [4]
Argentina	299,664	268,385	210,639 [5]
Mendoza	217,939	157,345	143,764 [6]
Tupungato	5,145	3,892	3,260
Tunuyán	3,901	2,499	2,262

Sources: 1) Estadísticas Agropecuarias, 2) Censo Nacional Agropecuario, 3) Instituto Nacional de Vitivinicultura, 4) Vinífera. p.8. Promedio 1986-1990., 5) Vinífera. Año 2. Nº Estraordinario, 1997, p. 8., 6 *Ibid*.

Table 6.2   Fruit tree area, in ha, 1988

Regions	Apple	Peach	Pear	Walnut	Cherry	Plum
Argentina	48,976	30,207	18,534	10,112	1,526	11,860
Mendoza	9,432	15,184	5,588	2,466	1,160	9,607
Tupungato	1,453	1,460	381	931	472	237
Tunuyán	5,112	791	1,171	1,091	264	283

Source: Censo Nacional Agropecuario, 1988.

Table 6.3   Fruit tree area, in ha, 1996

Regions	Apple	Peach	Pear	Walnut	Cherry	Plum
Mendoza	11,263	20,531	7,378	2,629	1,786	20,355
Tupungato	1,470	1,429	592	965	663	408
Tunuyán	6,356	2,052	2,044	1,197	438	551

Source: Actualización del Censo Frutícola Provincial, 1996.

As seen from tables above the region appears as highly diversified in comparison with other oasis areas where vineyards stretch over as a single crop. In addition these two departments differentiate from each other by fruit tree specialisation in Tunuyán. The evolution of the various forms of land cultivation in the last twenty five years shows a retraction of the vineyard area due to few reasons:

- The improvident expansion of the area devoted to common grapes;
- The diminishing of wine consumption from 90 to 40 litres per inhabitant between 1970 and 1996, in line with what happened in other countries with a wine tradition;
- The areas devoted to fruit trees and garden produce have increased following a dynamics stimulated by the economic re-conversion measures of the provincial government, the producers equilibrium strategies, and
- The incentives to international trade, as is the case with garlic.

These advances and regressions have been a constant feature in the region.

Argentine and Mendoza's economic policies

In the last four years a new process, as a result of public and private policies implemented in Argentina to incorporate the country to the global market has been observed. In saying this we are making reference to the convertibility law and to the foreign investment incentives established by the nation. Special loans have been granted in the province to redirect production. The Instituto Nacional de Vitivinicultura has incorporated active promotion functions. Of course, these are not the only measures taken, but certainly the most important. With the dollar peso even exchange system the Convertibility Law put an end to the inflationary process that during decades had plagued Argentine economy, and attained a structural transformation in prices stability. Next to convertibility, foreign investments were made easier by a decree of the Presidency in 1991 whereby Argentina joined the World Bank Constitutional Agreement of the Investment Warrants Multilateral Organism to ban all restrictions to foreign enterprise operations. In 1993 the Foreign Investment law does away with all requirements for dividend remission and capital repatriation, freeing them from any specific tax (Ferrer, 1998, pp.89-105), such as the obligation to reinvest utilities.

On its part the province of Mendoza has created the Transformation and Development Fund with money coming from privatisation carried out by the Nation and by the province itself. Loans to incorporate innovations to economic activities figure among the various destinations foreseen for this capital. At the sector level the Instituto Nacional de Vitivinicultura (INV) organises and participates in international conferences, and divulges the agronomic and enological requisites as well as the strategy indispensable for the incorporation to the wine and grape world market.

Private attitude responds to the pressure excerted by globalisation and to the export incentives established by the economic policy. In this regard let us add that all the major companies in the province have started their way towards an adaptation to the world market, and that other companies of foreign origin, obeying to the same steps and other particular incentives have joined them.

The ecosystem and its occupation

It is in the superior Tunuyán river basin where our study area lies, whose precise location from the administrative point of view is a portion of three districts: Gualtallary in Tupungato, Los Arboles and Los Chacayes in Tunuyán, covering an area of 900 km^2 (Figure 6.2). The piedmont, already mentioned, is part of an ecological system, and is a two level accumulation glacis – of which the lower one is more significant, newer, and broader – produced by arid and semiarid zone morphogenetic processes with torrential rains alternating with dry periods.

Long north-south fault lines separate the two levels, and also separate the piedmont from the Cordillera of the Andes that at this point reaches an altitude of 6,000 meters, and from an accumulation plain to the East (Barrera, 1968, pp.108-133). The glacis is entirely covered with deep layers of alluvial matter, with a slope between 0% and 89%, with a media of 6.2% approximately. At this point the glacis reaches its maximum West-East extension – 23 km and climbs from an altitude of 1,250 m in the East to an altitude of 2,250 m in the West. The river Las Tunas that flows all year around is born in the heart of the Cordillera and runs from West to East; its fluvial bed shows two terrace levels. Other watercourses born on the piedmont run only sporadically. In addition, the piedmont is part of the feeding area of a huge underground natural cistern with water of excellent quality. There is a high-pressure confined aquifer at the depth of 250 m.

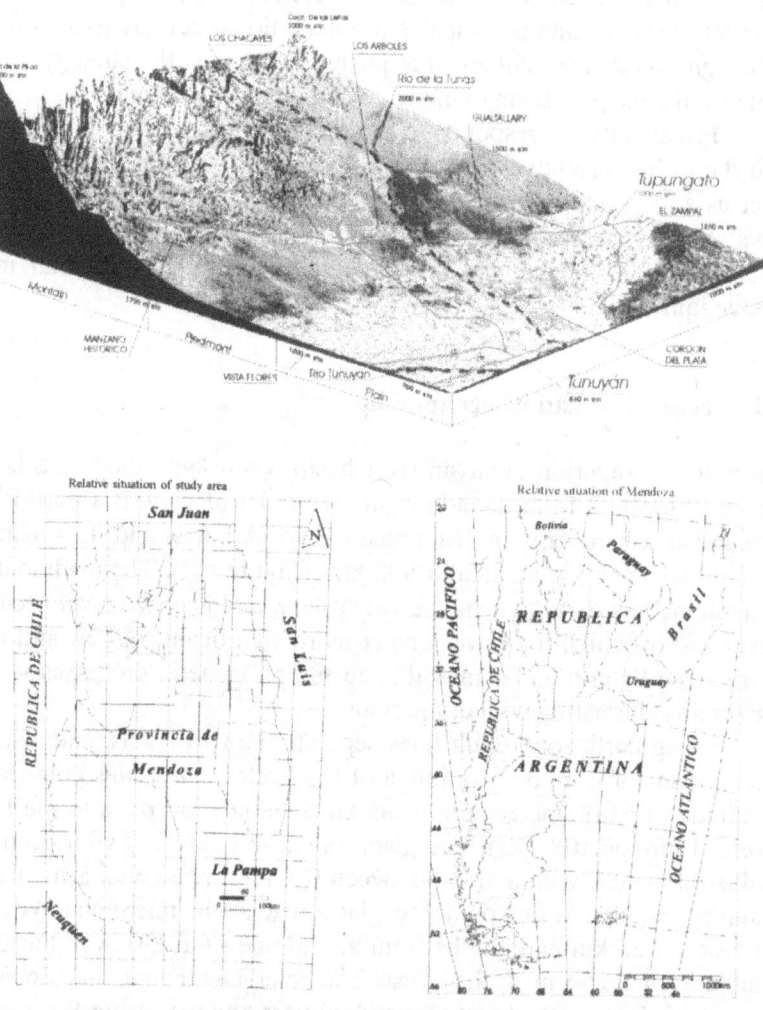

Figure 6.2   **The study area: Uco Valley, Mendoza, Argentina**

The dominating soils are a combination of typical 'torrifluventes' with very stony 'torriortentes', with limitations due to slope, stone abundance, and variable salinity (INTA, Project PNUD Argentina, 1985, p.100).

The climate is tempered semiarid, with an annual highest temperature of 21°C and a lowest of 5°C, and a rainfall of 227 mm. It is free from violent winds, and yet is affected by a foehn type wind, dry and non-periodic. As a result from the topography, soil, and climate, the vegetation is that of a *larrea divaricata* shrubby steppe with *stipa tenuissima* pastures.

During 400 years this ecosystem has suffered little modification from an extensive cattle raising activity tied to economic and climatic cycles (Prieto and Abraham, 1993-1994). Early, in the XVII and XVIII centuries, it was an area for raising and fattening cattle to be sold in Chile, and later was a fattening area of cattle coming from the humid pampa, with the same destination, Chile. Local factors: changes towards a model based on the vine and wine culture for the internal market; then national factors: the agro-exporting model; and international factors: Chile's direct trade with the Pampean region, provoked the decay of those early economic circuits. Thus cattle raising in Mendoza, and therefore in Gualtallary, Los Arboles, and Los Chacayes went their way towards marginalization. Up to that moment the large estates in the piedmont kept extensive cattle raising without adding value to the land, while the immediate plain was progressively transforming itself as the result of an intensive agricultural activity dealing with permanent crops under irrigation. In the piedmont, topography was a limiting factor against that type of agricultural advances since the available irrigation techniques at the time allow only the incorporation of levelled lots. Thus it remained as an agrarian frontier up to practically today when agriculture in Mendoza, especially the vine and wine industry, enters into a process of economic globalisation.

In a word, the step separating the piedmont from the plain has been a fixation line, a level beyond which agriculture has progressed linearly in some sectors with fruit trees and vineyards, while the rest, when not idle, remained under a weak and extensive cattle-raising activity. At the moment, however, we are witnessing the incorporation of these lands around Gualtallary, Los Arboles, and Los Chacayes, the broader part of the piedmont. This was made possible by the area aptitude for the production of high quality wine grapes, that the elaboration of export wines requires, such as Malbec, Merlot, Cabernet-Sauvignon and Bonarda among the red ones, and Semillón, Chardonnay and Chenin among the white. This aptitude is the result of a stony and very permeable soil, with neither drainage nor salinity problems. Together with an annual and daily thermal amplitude (15°C the latter) it gives highly favourable conditions for the production of grape colour and tannin that quality wines demand, with a yield of 130 metric quintals per hectare (Vinífera, 1997, p.32).

Thus the opening of argentine vine and wine industry to the world market, the national and provincial economic measures, and the environment aptitude attracted the attention of local and foreign companies with large investment capacity. Investments indispensable for soil preparation, extraction of ground water, installation of water drip irrigation systems, construction of vine conduction and support structures, acquisition of fine vine varieties, agronomic technologies such as plastic tubes to stimulate growth. Attention must also be given to the mechanisation of cultural labours – harvest and grafting – in addition to housing, deposits, and, in some cases, to the construction of wineries and nurseries.

Among the local companies involved in this transformation process there are some large integrated wineries with experience in the production of quality wines and in their exportation, among others Catena, Reina Rutini, Flichman, Basso, Pulenta, Rubino, while in with foreign capital figure some Chilean, French, Dutch, American, Portuguese, and Spanish firms. Some bought virgin lands, and others already organised lots and abandoned properties that were enlarged by successive additions. A look at the landscape soon shows the various stages of an accelerated transformation, with machines preparing the ground, lines being laid for water dripping irrigation, posts being planted for vineyard infrastructure, path tracing, construction sites, that are all signs of an agro-industrial development.

Conclusions

In fact, the purpose of all these efforts is to reach the highest possible quality levels by transforming the vine and wine culture so that Argentine may become one of the leading countries in the industry. This means a positive attitude to face the state of crisis, and confidence in its possibilities. The visible signs are spectacular and anticipate the investment success. However, there is a high uncertainty level, in view of the trajectory of agriculture in Mendoza, the country's economic stability, and entrepreneurial continuity. With regard to the first point let us remark that the history of vine and wine in Mendoza is characterised by a permanent state of crisis apparent in the surging and disappearance of industrial plants, abandoned vineyards and wineries, eradicated vineyards, etc. Also this new boom has some important weak points, and the most critical of them is the dependency on a very competitive market to which Argentina is a recent newcomer. In addition, its success appears as closely tied to the continuation of the country's official economic policy that wholly warrants all foreign investments and supports all exporting initiative (see Figure 6.3). As to the attitude of the enterprises, our opinion, in view of the previous history of the involved firms, is that the expansion of local capital is a solid expression of economic dynamism. Instead the permanency of foreign companies appears as much more uncertain. Our perception sees a difference between Chilean enterprises and the others. Chilean companies push their roots into a market reached long time ago and are venturing in an environment similar but less risky than their own in Chile. As to the other countries their venture is much more doubtful, except perhaps that of a firm of French origin with an important trajectory in the province, Chandon.

Although the above mentioned agro-industrial facts may have a positive effect in the provincial vine and wine industry, the balance of the consequences on the local media posses many questions, be it as to the degradation of the natural environment or as to their insertion into the social context (see Figure 6.3). The area was already defined as a fragile ecosystem with occupational risks. Among the weaknesses we pointed out; the two most serious problems are related to underground water and slope. Water is extracted from a little known aquifer. The studies are of general nature and on the basis of that evaluation it is difficult to foresee the resource recuperation capacity, water table contamination, and hydrological balance between the underground water used in the piedmont and that consumed in the plain. On the other hand, it is well known that agricultural usage of areas with a marked slope produces a negative impact due to erosion processes and diminution of slope stability as a consequence of the disappearance of the vegetation covering. This last difficulty is further aggravated by the fact that slope cultivation is not an usual practice in Mendoza, the piedmont basin of the superior Tunuyán river being the only area showing some advances in this sense, with some forestry, fruit tree orchards, and itinerant horticulture. To these risks still another should be added: the torrentiality of watercourses produced by summer rains concentrated in time and space.

As to the insertion into the social environment the change initiation lapse is much too brief to risk a definite judgement. But our experience with the relations between a large enterprise and settlements in this century seems to indicate that we are in the presence of a model different from the one that first appeared at the beginning of this century and another in the sixties. The former gave life to towns around the wineries, while the second produced dwelling areas for the employees within the property limits. In the new model instead we noticed less number of employees and transitory labour relations, and consequently no relation with the population process.

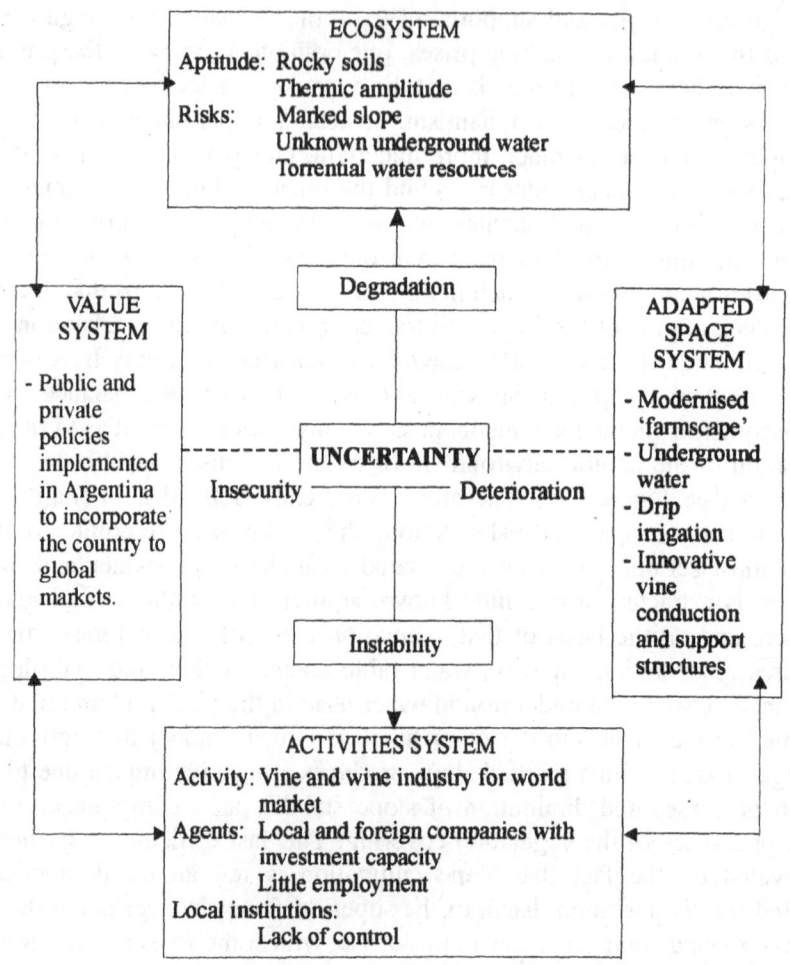

Figure 6.3   A systemic approach toward vine and wine industry

This assertion follows the verification of the absence of dwellings recently built or under construction, along with the intentions, revealed by the surveys, of a great mechanisation of all productive stages, and thereby little work force. Thus employment will only concern skilled labour-force such engineers, technicians, etc.

On the other hand, among the many relations with other local companies the possibility of divulging the requirements of the exporting mode can be perceived, as well as the results of new irrigation and cultural technical advances. Anyhow, due to their low investment capacity, few local producers have access to these new proposals.

This uncertainty grows if we take into account the vineyards abandoned after the agricultural expansion that took place in the sixties and seventies. Would this new colonisation have similar results?

Some effects of the new investments are perceived in the nearest city, Tupungato. Real state business of old vineyards and virgin lands have been reactivated. But the local authority, the municipality, has no control in the new enterprises. And this absence of local control appears also in other aspects, such as lack of environment protection, production planning, reinvestments under the form of services, and the already mentioned tax exception for foreign investment. All these factors support our uncertainty, this uncertainty also applies to other areas in the country were similar situations occur, even though the aim, size, and milieu may change.

References

Barrera, R. (1968), 'Geografía General y Geografía Regional', in M. Zamorano, La *Geografía en la República Argentina*, Paidós, Buenos Aires.

Ferrer, A. (1998), El capitalismo argentino, F.C.E., Buenos Aires.

INTA, Proyecto PNUD ARG 85/19. Area edafológica (c. 1990), *Atlas de Suelos de la República Argentina*, Provincia de Mendoza, Mendoza.

INV (1997), Vinífera, Año II, No Extraordinario.

Furlani de Civit, M.E., Gutiérrez de Manchón, M.J. and Pérez Romagnoli E. (1994), *Agroindustrial Transformation Model in Mendoza, Argentina*, Proceedings of the International Geographical Union Regional Conference, CD-ROM, Prague.

Furlani de Civit, M.E. and M.J. Gutiérrez de Manchón (1994), 'Possibilities of Endogenous Development in Minor Center of Mendoza, Argentina', in Chang-Yi David Chang (ed.), *Proceedings of the Study Group on Development Issues in Marginal Regions*, National Taiwan University, Taipei.

Furlani de Civit, M.E., Gutiérrez de Manchón, M.J., García de Martín, G. and Pérez Romagnoli, E. (1996), 'Relaciones entre un municipio deprimido y una agroindustria innovadora', in M.E. Furlani de Civit, and M.J. Gutiérrez de Manchón, (eds.), *Mendoza: Una geografía en transformación*, Facultad de Filosofía y Letras, Mendoza.

Furlani de Civit, M.E. and Gutiérrez de Manchón, M.J. (1996), 'Dinámica agraria en un oasis de especialización vitícola', in M.E. Furlani de Civit and M.J. Gutiérrez de Manchón, (eds.), *Mendoza: Una geografía en transformación*, Facultad de Filosofía y Letras, Mendoza.

Prieto, M. and Abraham, E. (1993-94), 'Proceso de ocupación del espacio y uso de los recursos en la vertiente nororiental de los Andes Centrales argentino-chilenos', *Cuadernos Geográficos*, 22-23, pp.219-238.

7 A place in globalization: Land use in the Argentinean Pampa

STELLA MARIS SHMITE DE CASTELL

Introduction

The globalization process shows an increasing interdependence in the world market, as well as the reorganisation of the production and commercialisation of agrarian products. The new tendencies of farming activities organisation and dynamic on a world scale and their repercussions in Argentina (see Teubal, 1995), also impact the provincial space.

As a result of the opportunities and/or restrictions of the global market, a new productive organisation model is being gestated in local space. This model has peculiar characteristics, closely tied to the natural conditions and, at the same time, directly related to the international market of agrarian products. The time-space evolution has brought about an agricultural production system orientated towards a predominantly overseas market, located in the provincial Northeast and East-centre, and constituted a marginal area within the Argentine pampean fertile plain. It is in this area, called 'espacio agropecuario de mercado' (market 'agropecuarian' space), where the most important activities are carried out and also where the intensification of the agrarian processes threats the preservation of soil quality. The rest of the territory, the 'espacio pastoril de subsistencia' (subsistence pastoral space), has not been incorporated to the extensive 'agropecuarian' production and remains out of the dynamic of the grain and meat national and international market.

The current world situation requires that La Pampa agricultural producers seek for commercial insertion ways that are increasingly competitive. Social specificities and productive peculiarities of local space are changing. It can be stated that the productive structure and social relations, under the impact of merging global network processes, internalise the changes and acquire a particular configuration. The articulation between what is local and what is global, determines the productive orientation and the development rate of productive activities as alterations are produced in the traditional social structure. New actors with a different productive logic intervene in the agrarian system, generating changes with a high degree of dynamism and uncertainty due to the consequences in the socio-economic structure of rural communities.

Importance of rural space

Tertiary activity accounts for the higher share in the Gross Regional Product (GRP) of La Pampa; primary activity, which consists basically of agriculture and cattle raising, comes next in importance. The composition of the GRP, according to 1993 statistic data, shows the following distribution; primary sector, 38.87%, secondary sector, 12.54%, and tertiary sector, 48.58%.

Within the agricultural GRP, farming and cattle raising represent nearly 100% (farming 40.4% and cattle raising 59.3%). It is interesting to note the evolutionary behaviour of the agricultural GRP, since the relation agriculture-cattle raising has been evidently modified in the last decades, as Table 7.1 shows (Dirección de ...). Agricultural activity increased during the 1980´s, it grew from 23.4% in 1980 to 40.4% in 1990; this trend has been stressed even more, in the 1990´s. There is a reduction of bovine cattle stocks that goes together with a decreasing importance of cattle raising in the GRP. According to the National Institute of Statistics and Census, Agrarian National Inquiry (INDEC), between the years 1995 and 1996, La Pampa was placed among the provinces that had the greatest bovine cattle stocks fall (-6.8%).

Table 7.1   Gross regional agrarian product by branches, (%)

Branches	1980	1981	1982	1983	1984	1985	1986	1987	1988	1989	1990
Agriculture	23.4	25.8	24.3	33.7	33.8	39.3	37.4	32.3	33.6	30.7	40.4
Cattle raising	76.3	73.9	75.4	66.1	65.9	60.4	62.3	67.3	65.9	68.9	59.3
Silviculture	0.3	0.3	0.3	0.2	0.3	0.3	0.3	0.4	0.5	0.4	0.3

Source: Our elaboration from CFI data (Federal Council of Investments).

To highlight the importance of the rural sector, another valid indicator is the Active Population - AP (Población Económicamente Activa - PEA). According to information given by the National Institute of Statistics and Census of the province of La Pampa, in 1991, 27.1% of the provincial AP was working in the primary sector, 18.2% in the secondary sector and

54.6% in the tertiary sector. The agrarian AP shows a falling tendency, the province of La Pampa has 36.1% less of rural population than three decades ago. According to INDEC data, rural population has decreased from 86,358 inhabitants in 1970 to 55,844 inhabitants in 1990. These changes in population distribution show an important migration process towards urban regions, especially towards the centres of major importance, a phenomenon already observed at national and even international scale. Although La Pampa has 80.8% of urban population, nevertheless it is placed among the provinces that account for the highest shares of national rural population. It is also situated amongst the provinces where the agrarian activities generate the largest exportable surplus. The share of agrarian goods in the provincial export goes up to 90.5%, of which approximately only a 10%, is exported with some added value.

Agriculture intensification

The agriculture system of the province of La Pampa is characterised by a combination of crop production and cattle raising. During the 1980's, the cattle-raising crisis impelled an important stock liquidation, whose recovery would take four or five years. The total area given over to crops is of around two million hectares, 50% of which falls to annual crops destined to overseas market (cereals and oilseed); besides, a further 25% of sown area is accounted by annual fodder for cattle feeding. Only the remaining 25% area under cultivation is dedicated to perennial pastureland. This means that the major part of the farming area is annually ploughed, increasing the risk of soil erosion. Thus, farming, especially in the case of second harvest crops (large grain in winter and small grain in summer), allows the achievement of short-term higher incomes, although this intensification does not meet the needs of sustainable farming. If this trend continues the problems of soil deterioration will increase (Morello *et al.*, 1997).

According to the table 7.2, it is possible to find two different types of annual:

- Small grain harvest, where wheat is the most important, and
- Large grain harvest, where sunflower and maize prevail.

These three major crops are used for grain production, while the rest of the crops could be utilised either for grain harvesting and/or as pasture accordingly to the system of agrarian production.

The provincial harvested area accounts for 12% of the whole area of the country and for 30.88% of the national cereals and oilseed production. The data related to surfaces and production volumes indicate a remarkable step forward in the farming process in the East departments of the province. This implies a more intensive use of lands with agronomic restrictions because of their marginal situation within the context of the pampean fertile plain.

Thus, despite the fact that the mixed model has proved to be less vulnerable to the economic circumstances and more adequate to the climatic characteristics of the region, this intensification process is transforming it, due to the low share given to fallow and perennial pasture intended to recover soil nutritional values of the arable layer.

Table 7.2   Cereals and oilseed production - La Pampa - 1995

Annual crops	Sown area (ha)	Harvested area (ha)	Provincial production (tons)	Provincial prod./national production (%)
Small grain harvest	441,651	422,219	521,872	9.09
Wheat	389,766	354,402	453,759	3.69
Oats	65,291	44,085	46,410	
Barley	7,949	7,949	11,651	* 5.40
Rye	31,645	15,683	10,052	
Large grain harvest	637,472	547,268	800,217	21.79
Oilseed	475,014	445,931	560,954	15.81
Maize	130,605	80,053	200,638	1.64
Sorghum grain	28,763	19,235	35,586	4.30
Soya	3,090	2,049	3,039	0.04

* Represents the percentage of oats, barley and rye participation

Source: National Agrarian Inquiry (ENA), 1995, INDEC, enlarged general results Vol. 5.

The intensification of farming activity, with stress on the double harvest (large grain harvest and small grain harvest) and cattle raising reduction, are related to the behaviour of the international market of agrarian goods. Table 7.3 shows the production of the three major crops, namely wheat, oilseed crops and maize, during the last seven agrarian campaigns. Up to now in the 1990's, the oilseed crop sown area has considerably increased, which is explained by a raising demand of oil and flour at world scale.

From the 1979/80 to the 1994/95 farming campaign, the oilseed crops sown area rose from 116,000 to 475,000 ha, while the production grew from 100,000 to 950,000 tons in the last fifteen years. On the contrary, the perennial pasture sown area (e.g., the alfalfa area), which is the one that mainly provides soil nourishment, only raised a 27%.

Table 7.3   Evolution of the production volumes of main crops (in tons)

Agrarian campaign	Wheat	Maize	Oilseed crops
1991/92	917,300	695,700	446,200
1992/93	857,000	538,100	386,300
1993/94	934,000	834,200	699,100
1994/95	661,000	232,000	793,500
1995/96	506,000	125,000	731,100
1996/97	715,000	360,000	835,500
1997/98	1,056,000	480,000	950,000

Sources: Storing cereals centre of La Pampa and bordering provinces and Secretariat of Agriculture, Cattle raising, Fishing and Feeding (SAGPYA).

In spite of the significant increase of oilseed crops, wheat is still the most important crop, especially for the south-central sector of the agrarian space of market. The largest oilseed crops sown area is located in the northeast departments of territory.

Agrarian management

Farming intensification is parallel to the increase of capital participation in the local productive circuit. The farming technological package applied in previous decades proved that agriculture is profitable with the contribution of intensive capital. This generated an accelerated change that manifests in the investment of national and foreign capital, coming in many cases from other productive sectors. Currently, the agrarian sector is profiled with a more dynamic role in the articulation of economic, social and political variables at the provincial as well as at the national level, but there are other agents: the new actors are the agrarian managers (Barsky *et al.*, 1997).

Due to the dualism caused by differential in capitalisation and technological incorporation, two well-separated components can be identified in the agrarian structure:

1. Managerial producers, with possibilities of productive development in competitive conditions, whose farms are located in the departments with the most fertile land, that is to say, where the highest productivity indexes are achieved.
2. Traditional producers, with worse productive resources, in some cases located in ecologically critical agricultural areas, with less decision power and increasing impoverishment.

Managerial producers are consolidating their hegemony, while at the same time they exercise a strong pressure on soil utilisation. Big owners or investors with greater financial assets demand land to them in order to implement a more intensive production. At the same time land supply consists basically of small plots, that do not make a profitable economic unit and whose owners are greatly in debt or have no technical or economic chances of achieving a sufficient level of production. This modality that is being introduced in a growing way, moves the productive activity that was in charge of the agrarian producer and his family, to the contractor or new owner (Barsky *et al.*, 1991, pp.662). This process implies a major scale productive enterprising, so that production becomes a 'farming business', whose primary aim is to get the highest benefit. In this framework, the traditional producer is unable to continue in his activity and sells or leases his property. However, when leasing, the previous investments done to his property (water bearers, wire fences, silos and housing) generally lose their value, to what the risk of inadequate soil use is added.

Farming by contract has become a profitable business and it attracts capital coming from non-farming sectors. New productive agents have little concern about soil preservation. They are not willing to protect it with conservative agronomic practices that raise costs, make the working period longer and require more manpower. This capital-intensive management is causing the substitution of the mix farming-cattle raising system by a system where cereals and oilseed crops predominate, and where technologies that increase per-unit yield are used.

It can be verified also that there is a resurgence of new ways of leasing lands, indeed very different to the ones implemented during the 1920's and 1930's. The 'new leasings' are carried out by contractors, cropping pools or

Farming Funds of Direct Investment (FAID), who contribute machinery and input.

This farming activity boom has caused an over-valuing of the lands of best aptitude (departments of Maracó, Chapaleufú, Conhelo y Quemú-Quemú), located in the East of the provincial territory, which are the most attractive area for cropping pools. Seven cropping pools have been operating in the last three years with a total investment of 35 million dollars. For the present year, the International Finance Corporation, a Farming Fund of Direct Investment (FAID), intends to start a total investment of 20 million dollars that will extend for seven years.

These processes of capital investments are accompanied by changes in the land tenure regime. It is becoming a model of space occupation and the kind of productive organisation of each agrarian establishment. These are the alterations that form a productive area with a complex and heterogeneous structuring. Traditional producers, who are dependent on capital fund contributions and technology, prevail in a local space that is of interest for foreign investors. The arrival of these investors are cause the appearance of the 'new owners'. Owners, who are stressing a localised productive specialisation in determined areas, according to the eco-farming aptitude (wheat crops, oilseed crops, cattle raising hibernation, feed-lot) as well as the beginning of non-traditional space uses (deer breeding, aromatic herbs, agritourism, honey production) (Barsky *et al.*, 1988, pp.326).

Conclusion

The impact of global processes in the agrarian market space in La Pampa province, can be summarised in two basic trends. On the one hand, changes are proved in the agrarian structures that are translated in a major productive transformation. The extensive farming-cattle raising mix model is changing into a productive specialisation based on few major crops. On the other hand, these productive changes are accompanied by transformations in the administrative management and organisation of agrarian establishments. Medium and small familiar exploitations are giving place to individual or societal enterprises that are located in the areas of greatest productivity.

The changes that are taking place in the rural community imply the research of increasing short-term profits, without taking into account if such behaviour is compatible with a sustainable management of soils. The achievement of economies of scale, with the intensification of the land productive management and the utilisation of agro-chemicals, has as a consequence an inadequate use of soils from a 'sustainability' viewpoint. The eco-farming damage is increasing and the survival of small and medium producers in the productive circuit is at risk.

References

Barsky, O. et al. (1988), *La agricultura pampeana, Transformaciones productivas y sociales*, Fondo de Cultura Económica, Buenos Aires.

Barsky, O. et al. (1991), *El desarrollo agropecuario pampeano*, INDEC, INTA, IICA, Buenos Aires.

Barsky, O. et al. (1997), *El agro pampeano. El fin de un período*, FLACSO, Oficina de publicaciones del CBC, Universidad de Buenos Aires, Buenos Aires.

Dirección de estadísticas y censos, Provincia de La Pampa, diversas publicaciones, INTA (Instituto Nacional de Tecnología Agropecuaria).

Morello, J., Marchetti, B. Rodriguez, A. and Nussbaum, A. (1997), *El ajuste estructural argentino y los cuatro jinetes del apocalipsis ambiental, Erosión del suelo, deforestación, pérdida de biodiversidad y contaminaciónhídrica*, Oficina de publicaciones del CBC, Universidad de Buenos Aires, Buenos Aires.

Teubal, M. (1995), *Globalización y expansión agroindustrial, ¿Superación de la pobreza en América Latina?* Editorial Corregidor, Buenos Aires.

8 The globalization impact on a marginal agricultural area

GLADYS MABEL TOURN

Introduction

This article deals with the problem of the impact of globalization on a marginal agricultural area in the province of La Pampa, Argentina and, among its main characteristics, aims to explain:

1. The influence of technological change on the information flows as well as on the findings destined to production improvement.
2. The increase of direct private investments on productive systems.

The present assertions have a high degree of generalisation due to two factors: As far as units of production is concerned, the condition of the farming sector is one of great heterogeneity. There are about 8,500 'farms' which range from 5,000 to 20,000 hectares, placed in areas of great diversity because of its farming qualities, what leads to deep differences between the production systems, the products obtained, and the know-how used. Broadly, they can be classified into three groups: the small domestic exploitation, with diversified production; the medium-sized property, whose size is similar to that of one economic unit; and the *estancia,* or large property, with predominant cattle-raising or mixed production. In addition, as it is a contemporary process, there are neither enough measurements, nor added or systematised information; therefore, the more evident processes will be derived from existent statistical records such as specialised articles and interviews to qualified informants.

19th century globalization

Aldo Ferrer (1997) holds the idea that globalization is not an original process. In the past, there took place a number of events that exerted a strong impact on the countries that shared the world order. As regards our region he expressed '...railway and steam. The technological improvement of transport during the 19th century produced a great reduction in freights and in the integration of isolated areas (as the *pampeana* region in Argentina) to the global system...'.

As a result of the new means of transportation and the occupation of new lands, between 1821 and 1915, nearly 50 million people – 90% from Europe – emigrated to the New World. This constitutes an extraordinary globalization phenomenon found at the very start of the development of colonisation in Argentina and other countries in America (Ferrer, 1997). The author believes that, in comparison with those processes, the present financial market integration and transnational corporation developments are less important facts (*Ibid*, p.37). The following data (Table 8.1) allow us to appraise the impact of globalization on our province in the mid 19th century.

Table 8.1   Province of La Pampa, comparison in 1895 and 1920

Indicators	Year 1895	Year 1920	Percentage
Population	25,914	122,535	+ 473
Urban settlements	4	53	+ 1,325
Railways (km)	241	1,526	+ 633
Cultivated land (ha)	3,695	1,506,716	+ 40,778

Sources: National Population Census 1895, Territorial Population Census 1920, Statistics.

Since territorial colonisation began, the railways reached their present-day limits, limits that coincided with those of the exploitation of natural resources. Twenty-five years later, colonisation had advanced up to the agricultural frontier boundary. Starting from an empty land after the displacement of indigenous peoples, space was organised in a way functional to the new world order established from Europe. In those years, the main characteristics of Argentina were configured: deep regional unbalances and floods of Europeans that altered the previous social and cultural patterns and constituted the prevailing basis of the country's present-day population.

20th century globalization

The present process is characterised by the impact of technological changes, especially on communications, on the amount of financial flows, on the increase of international trade, as well as on direct private investments by transnational corporations, and on the changes on the economic regulatory framework (Benko, 1995, pp.41-49). We will analyse the influence of some of these facts on the behaviour of some variables related to agricultural-and-livestock production, which contributes half of its gross domestic product in the province of La Pampa.

An inflexion point, the convertibility law

The fact that really marked a changing point in this activity was the Convertibility Law, sanctioned in 1991, that, among other measures, established the *peso*-American dollar parity and the prohibition of debt indexation, allowing since then the stoppage of the inflationary process. Economic stability introduced new working rules for all sectors, including the farming one. This, added to the suppression of a package of government regulatory measures, eliminated the elements that separated farmers from the working conditions of the global market and built a new production framework. From then on, international prices governed production as well as a great deal of inputs (Barsky *et al.*, 1997, p.5).

These measures found producers in differentiated situations; in broad outline, producers can be classified into two great groups: those without debts, that attempted to adapt to the new order with greater or lesser difficulties, and those that had been unable to overcome their critical situation. A farmer said: 'In the decade of 1990 we had to put ourselves to work hard because of stability, deregulation, market economy and globalization, that imposed competition on us' (*La Arena*, Interview), an expression that clearly depicts the new situation: either there is an adaptation to working under different conditions, or the exploitation is abandoned.

Access to information and incorporation of technology

The need of increasing scale production, of running of farms in an entrepreneurial style, and of increasing yields compelled farmers to adopt new technologies for greater efficiency (Caviglia, 1994, pp.19-25). Since the 1970's great transformations are being produced, specially in agriculture, that has notably increased yields: in the first place by the incorporation of hybrid seeds, as sorghum grain, that in the next decade will start to be replaced by sunflower. Simultaneously, Mexican disease-resistant varieties of wheat were incorporated.

In our decade, sunflower is the crop that has reached greater development, wheat fertilisation has originated greater productivity, and the use of herbicides and pesticides has been intensified. No-tillage seeding, on the other hand, will exert a great impact by the end of the decade. The result of these innovations can be viewed in Table 8.2.

Table 8.2   Agricultural production and yield (kg/ha)

Decade	Wheat	Maize	Sunflower
1980	1,365	2,422	1,192
1990	1,384	2,772	1,612

Sources: National Bureau of Agricultural-and-Livestock Technology (INTA), Secretariat of Agriculture, Cattle-raising, Fishing and Food Office (SAGPyA).

On the other hand, the cattle-raising sector has a slower growth. Its transformation demands the adoption of process and management technologies that are more abstract and, consequently, are adopted more slowly by farmers. One can ask now, how many producers have access to this information. To advice them, there are five established programs carried out by national and provincial bureau's. They systematically meet the needs of about 10% of the approximately 8,500 existing producers.

In the last years, the participation of professionals individually engaged in cases of adoption of technological packages that demand permanent advice was enlarged. Moreover, there are those who adopt innovations by merely observing their neighbours, that is, by demonstration effect. The universe is then composed by systematic or intuitive technology adopters, who can be classified into early, middle or late decision takers. We must also notice that not all those who receive advice are successful in their application. Then, between the available techniques and the results, there are long distances, determined by a myriad of different situations. With regards to the perception of space, while some people have a clear conscience of a wide world space, others are enclosed in their local scope, another factor that influences their decision taking.

Farmers that do not incorporate innovations find it increasingly difficult to remain within the productive system (Viglizzo *et al.*, 1994, p.99). Then they can leave the country and lease their property. Thus, they become partners in a major scale business. Their way of life becomes urban, they integrate another social framework; therefore, the social features of the traditional rural world turn to dilute.

The increase of direct private investments

One other evident manifestation of these changes, in this decade, is the arrival of extra-regional capitals in two ways: cropping pools and agrarian investment pools. These pools are located in the middle-east of the province and it is not possible to know exactly what area they exploit. One of them has projected to work 52,000 ha. in seven years, which is more or less 20% of the land sown with wheat, the main small grain harvest of the province. In general, there are Argentine and foreign capitalists, whose benefits are obtained from a combination of factors:

- They diminish climatic risk by working lands in different regions.
- They use advanced technological packages with the aim of increasing productivity.
- The volume of available financial resources allow them a wide scope of savings and profits, i.e., through advanced payment when leasing lands, through the purchasing of great inputs, and the selling of large outputs, all of which contribute to better trade prices.
- The production scale allows them to negotiate better purchasing conditions with the contractors, who are in charge of the farming labours.

Another characteristic of this style of exploitation is that they choose ecologically suitable farms and, as their goal is gaining the greatest possible profit, there are not enough precautions taken in keeping the sustainability of resources, basically the soil. Simultaneously, they do not establish regional bonds because they neither make permanent investments nor do they generate constant labour relations. On the other hand, there remain the previously mentioned traditional producers. Compared with big entrepreneurs, they have small parcels, high production costs, less information and, generally, a more conservative attitude; all this delays the adoption of changes. As a result, their efficiency level is lower and sometimes, as has been explained before, they prefer to emigrate. In this way, there appears the fragmentation of the productive system, which will keep being formed by producers of a great heterogeneity, with very different possibilities of permanence within an increasingly globalizing context.

The deepening of regional differences

In addition, these changes are localised in a narrow fringe of territory. The rest, especially when advancing towards the West, is not reached by any of those changes. Here the differences are accentuated not by endogenous changes but by the shift of its relative position within the whole: eastern dynamism emphasises the static characteristics of the West, where marginality deepens.

Comparing both regions, it is evident that in the latter there is little population, lack of information, reflected in the absence of technological advice, and an increase of poverty, all of which emphasise the precarious conditions of life. Therefore, far from reaching integration into one global economy, it seems still more difficult to leave the circle of extensive low profit cattle-raising, scarcity of population, poverty and isolation.

Conclusions

Focusing on the changes in the agricultural sector in the province of La Pampa, we can agree with Aldo Ferrer that these are far from producing the impact of the 19th century process. Although in one sense the transformation is similar, it is also true that the space-time framework has deep differences. Furthermore, while in the past century capitals searched areas with available natural resources, today they add to that condition the existence of an infrastructure that allows them immediate production, without previous investments. We observe trends towards fragmentation in the farmer's profile, with social and cultural consequences not yet evaluated. The great transformation is concentrated on a narrow fringe of the province, that deepens the regional unbalance.

References

Barsky, O. *et al.* (1997), *El agro pampeano. El fin de un Período,* FLACSO (Facultad Latinoamericana de Ciencias Sociales), Oficina de publicaciones del CBC, Universidad de Buenos Aires, Buenos Aires.

Benko, G. (1995), *Economía, Espaço e Globalizaçao na aurora do Século XXI*, Ed.Hucitec Ltda., Sao Paulo.

Caviglia, J. (1994), 'Capacitar para reconvertir: ¿Estrategia de extensión? De la transferencia de información técnica a la capacitación en la toma de decisiones', *Horizonte Agro-*

Económico, Instituto Nacional de Tecnología Agropecuaria. (INTA), 1989: 3, La Pampa-San Luis.

Ferrer, Aldo (1997), *Hechos y ficciones de la globalización, Argentina y el Mercosur en el sistema internacional,* Fondo de Cultura Económica S.A., Buenos Aires.

Viglizzo, E.F. *et al.* (1994), 'La reconversión de la empresa rural, Ficciones, realidades y alternativas', *Horizonte Agro-Económico.* Vol.1, N° 3. Instituto Nacional de Tecnología Agropecuaria (INTA), La Pampa-San Luis.

9 Transforming the fringe:
Tobacco-related wood usage and its environmental implications

HELMUT GEIST

Introduction

From the viewpoint of a global perspective of changes in land use and cover, Williams (1994a) noted that felling trees for the combined objectives of obtaining wood for construction, shelter and toolmaking, of providing fuel for commercial as well as domestic purposes, and of creating land for agriculture 'has culminated in one of the main processes whereby humankind has modified the world's surface of vegetation'. However, 'the distribution, quantitative extent, and rate of change in the area of forest, through both deforestation and reforestation, have been and remain subjects of great debate and uncertainty'.

This paper uses a crop-specific approach by addressing the notion of felling trees in the special case of tobacco. The farming of the crop seems to become increasingly understood as one of the major social and economic driving forces of global environmental change since about the start of commercial growing in early 17th century (Geist, 1998a, b; Spada and Scheuermann, 1998). The paper first outlines the significance of tobacco farming as it mainly occurs at the agriculture/forest frontier in growing countries of the developing world within the present context of efforts aimed at regulating the global tobacco industry. While historical data are hardly to come by, major emphasis is put upon empirically derived dimensions of recent wood usage for the curing of the crop during the first half of the 1990s. From this, the resulting consequences upon land use and land cover changes are indicatively derived.

The results gained will by large contrast the pattern of tobacco's environmental implications as outlined in recent tobacco industry commissioned reports, but also as perceived in research programmes or networks on global environmental change (and related land-use and land-cover changes). Particularly industry commissioned reports seem to repudiate previous acknowledgements of the problem by downsizing the special issue of deforestation due to curing of tobacco.

Tobacco's part in transforming the 'fringe'

Transformation processes at the fringe

By relating to the notion of margin (or 'fringe'), uses are made of at least three conceptual designs and observations as outlined. First, and according to the agricultural location theory of von Thünen, tobacco could be set up as one of the few crops still suitable to be grown in the last (concentric) ring of land use around a central town and in close distance (or even encroaching upon) the heavily forested 'cultivable wilderness'. This is mainly due to low transportation costs (versus high value) and low perishability of the cured crop (O'Kelly and Bryan, 1996).

Secondly, and in historical as well as recent terms, the destructive exploitation or economic plunder of natural resources that came about with the diffusion of modern (agro)capitalism and its movement overseas can be identified as a paramount driving force under the heading of ('non-romanticised') frontier or colonisation studies (Williams, 1994b).

Thirdly, from its biological properties and inner limits of growth, tobacco can be cultivated where a long enough dry season allowing for harvesting and curing the crop, sufficient rainfall (or irrigation) and frost-free days are available, i.e., not so under arid, humid or polar, but under semiarid to semi-humid climates in the tropical and subtropical (as well as temperate) zones (Andreae, 1981; Akehurst, 1981).

Fourthly, with only few exceptions (such as the low-lying growing areas of Andrah Pradesh in India or the farming of cigar tobacco in valley bottoms of insular South East Asia), tobacco growing mainly occurs in hilly or highland areas of the developing world (Akehurst, 1981). Thus, the natural environments where tobacco is commonly grown could be characterised as upland dry forests and woodlands being seen as the 'most endangered major tropical ecosystem' (Janzen, 1988).

Fifthly, since tobacco is a very labour-intensive crop that needs large amounts of (mainly seasonal) workforce to be organised during the agricultural cycle, the bulk of tobacco has shifted into the developing world where conditions of low cost production are given, particularly in the form of (unpaid) women and children labour. As compared to the 1700-1800s, when most of global tobacco had been concentrated in northern hemisphere America (such as Chesapeake Bay area and the Caribbean), with the breakdown of colonial rule (e.g., slavery and indentured labour) following the American Revolution, tobacco farming spread into more than 120 growing countries at present (Goodman, 1995).

Within almost a century's time, world tobacco production increased from about 2 million tons in 1909/13 to nearly 3 million tons in 1948/50, and, stimulated by rising demand, further inclined to about 7.5 million tons in the 1990s. In Table 9.1, data prove that at present about 80 percent of global tobacco is produced in growing countries of the developing world (with most of the tobacco originating from low income countries of the tropical and subtropical zones in Africa, Latin America and, particularly, in Asia).

Table 9.1   World tobacco production, 1961 to 1998 (3-years' averages, for 1997/98 2-year average)

	Total Production (metric tons)	Developing countries as % of total produce	Share of Africa
1961/63	3,874,267	51.1	5.2%
1964/66	4,521,399	54.0	5.5%
1967/69	4,757,865	57.9	4.5%
1970/72	4,673,295	58.2	3.8%
1973/75	5,202,918	58.3	4.5%
1976/78	5,731,196	60.8	4.8%
1979/81	5,773,452	61.7	5.0%
1982/84	6,162,715	70.2	4.8%
1985/87	6,403,185	69.4	4.8%
1988/90	6,971,452	74.5	4.9%
1991/93	7,753,329	81.0	6.0%
1994/96	6,721,548	79.0	6.7%
1997/98*	7,770,217	79.9	6.5%

Source: Food and Agriculture Organization of the United Nations (1990-98), *Internet Database of Primary Crop Production*, FAO, Rome (http://www.fao.org).

Towards regulating a global industry

Except for government interventions (such as tax policy) and the most recent litigation of the U.S. cigarette industry, the tobacco sector had so far no major regulations to undergo. It could be considered an almost completely deregulated sector in the world economy, particularly due to the unique combination of 'policy networks and the cigarette industry' (Read, 1996). The market for tobacco leaf, unlike coffee, tea and sugar, does not follow a structuration by production quotas, and, thus, the crop comes close to be a free market crop. However, from the methods of commercial operations, or what Tucker (1982) called the 'tobacco cycle', it is useful to make a distinction between the growing and curing of green leaf tobacco, buying cured leaf and the processing as well as manufacturing of tobacco products (with institutional interlinkages possible between the operations). Other tobacco related activities are distribution, transportation, advertising, sale, paper, printing and insurance (however, not considered here).

The farming sector turns out to be characterised by mostly small farms under the peasant mode of production with average sizes internationally covering between 0.5 and 1.5 hectares of land. Due to its high labour intensity, the crop has proved to be rather resistant towards large-scale production and mechanisation, while this nevertheless implies that 'farms are usually owned by large national growers' (Tucker, 1982), be it under nationalised monopolies, individual holdings or (foreign) transnational agro-businesses. The notion exists that in many growing countries (particularly of the developing world) the political and economic importance of farming constitutes a major barrier of diversification or shifting to alternative crops (UNCTAD, 1995; Altman, *et al.*, 1996).

In buying, either done for a government buying agency or a private leaf supplier, there are two general mechanisms at work. In the developing world, contract farming is the main operation with buyers arriving at the farm or buying stations set up after harvest (less often, the latter could be independent selling places if there is no central agency). In developed countries, the prevailing mechanism is the tobacco auction, however, also practised in Malawi and Zimbabwe. The buyers could be domestic manufacturers, foreign manufacturer's buyers, and international as well as local merchants (Tucker, 1982).

The sector of tobacco processing and cigarette manufacture, being the main stakeholders in tobacco, is characterised by an oligopolistic structure with a few transnational corporations (TNCs) having enlarged their global operations from about one third of the world market in 1980 to about 90% since 1990 (Pater, 1994). This could mainly be seen as a result of the combined effects of heavy spending on advertising/sponsorship and structural adjustment measures/breakdown of national tobacco monopolies (as well as national companies) since the 1980s, i.e., buying newly privatised cigarette companies, setting up joint ventures and building distribution and sales networks mainly in emerging markets of Eastern Europe and the developing world (but also to locate manufacturing close to the sources of tobacco leaf increasingly sold overseas). The leading, mostly North American or European companies with truly multi/transnational operations are Philip Morris, British American Tobacco (BAT), Reynolds, Rothmans, Japan Tobacco Industries (JTI) and Reemtsma. From the bulk of turnover, the largest company still is Chinese National Tobacco Corporation (CNTC), however, operating in an up to now closed national economy. As compared to the 1950s, when the companies started to diversify, e.g. into paper and packaging, energy, retailing, industrial and fast-moving consumer goods, a reversal of the trend could be assumed from the most recent merger of BAT and Rothmans in 1999.[1] At present, and to shield their assets from lawsuits in developed countries, TNCs are found to shift markets increasingly to developing countries. With, currently, Philip Morris, BAT and Reynolds each owning or leasing plants in at least 50 different countries of the world, the 'overseas expansion has transformed the tobacco industry around the globe', as stated by Hammond (1998). He further notes that with the inflow of particularly U.S. companies into countries such as South Korea, Thailand and Taiwan smoking rates rise after these countries were forced to open their markets to U.S. tobacco exports.

Mostly for reasons of public health (and on the basis that epidemiological evidence on smoking has improved over the last 40 years), the introduction of regulatory measures is considered to control the growing 'tobacco epidemic' particularly in the developing world. On the background of (good) 'global governance', Abedian (1998) designed a framework of tobacco control policy instruments that, if applied, will touch next to all operational fields of the industry as outlined. With some of the instruments already used, they aim at 'demand-side instruments' (such as taxation, advertising and sponsorship regulation) as well as at 'supply-side instruments' (such as substitute crop development and trade-related issues). Currently, efforts are mainly directed towards governments and cigarette manufacturers being 'some of the richest, most powerful and politically influential transnational companies in the world ... (that) remain securely attached to a product which is cheap to make and highly profitable', and, 'being addictive, also easy to sell' (Read, 1996). Since the impetus, however, of the 'Bellagio Statement on Tobacco and Sustainable Development',[2] tobacco control policies have gained a special momentum out of social, economic and even environmental reasons.

Tobacco's critical impact upon the environment

Goodland, *et al.* (1984) were among the first to raise the issue of tobacco's critical impact upon the environment in the developing world. With special regard to tropical ecosystems they stated that the 'relationships among tobacco curing, fuelwood shortages, and deforestation and other forms of environmental degradation are becoming increasingly clear'. Tobacco's impact was seen to fall into three broad categories, i.e. 'soil degradation', usage of 'biocides', and 'fuelwood crisis'.

By 'soil degradation', reference is commonly made to tobacco's large demand of macronutrients from the soil (such as nitrogen, phosphorus and, particularly, potassium). Harvested as an annually grown crop and usually restricted to one crop in three or four years, tobacco requires either fertile soils or regular inputs of fertiliser. Given the generally low nutrient content of tropical soils, criticality emerges in that 'tobacco depletes soil nutrients at a much higher rate than many other crops, thus rapidly decreasing the life of the soil' (Goodland, *et al.*, 1984). As from the societal (and technological) responses of tobacco farmers to environmental change, it could be stated that an alternative to dependence on commercial fertiliser is to exhaust soil fertility and then to clear new land (what historically had often been the case by deforestation and is occasionally practised still today) or, after mining the soils for sufficiently enough 'brown gold', to shift to non-agricultural uses ('de-agrarianisation') (Geist, 1999a).

The usage of 'biocides', i.e. herbicides, nematocides, fungicides, insecticides and other chemicals, not only bears the possibility of crop contamination either with inherent danger to those who smoke or chew the leaf or with occupational hazards for farmers and their families, but also relates to the contamination of land and water-supply, thus contributing to soil (or land) degradation. Tobacco is considered to be 'one of the crops on which biocides are used most heavily ... (with) vast quantities ... applied ... virtually throughout its seven to eight month growing season' (Goodland, *et al.*, 1984). Among the registered active (1,500) and inert (900) ingredients, around 150 different types are used in tobacco farming with a maximum of 16 applications during one growing cycle. Some of the biocides are toxic or carcinogenic, and, though mostly banned or severely restricted in the developed world (such as aldrin, DDT and lindan), still used in marginal areas of the developing world (Pater, 1994). Interlinkages with soil nutrient depletion could be seen in that rapid soil fertility decline needs artificial inputs (besides fertiliser) to protect plant growth and avoid crop pests to become endemic.

Widespread usage of wood

The tobacco industry, i.e. farming, cigarette manufacture and packaging for export, requires substantial amounts of wood for a variety of purposes. Wood in farming is principally used as fuel for curing[3] (fuelwood, firewood), but smaller amounts are also directly used in the form of poles and sticks in barn construction (polewood). It has further to be noted that 'substantial quantities are used indirectly in a wide range of paper and paperboard products required for cigarette manufacture and packaging' (Fraser, 1986), with such usage, however, not considered here.

Sources of energy used

Fire-cured tobacco In the 1990s, (dark) fire-cured tobaccos were grown in about 20 producer countries, most of them in Africa (USDA/FAS, 1994, 1997). The sources of energy used to produce smoke are exclusively solid wood or by-products from trees, e.g. harvested by large mills such as slabs and sawdust. In addition, cases are reported of supportive fire-curing also in naturally cured tobaccos, e.g., with Japanese air-cured types, 'Jaffna-cured' chewing leaf, and wrapper cured Sumatra tobacco (Akehurst, 1981). These cases, however, are not considered here (Table 9.2).

Table 9.2   Curing fuel as a percentage of flue tobacco cured, 1993

Region	Coal	Oil/Gas	Wood
North & Central America		98.62	0.40
South America	0.28	6.57	93.16
Europe (incl. Turkey)	6.16	93.15	0.68
Africa	72.93		27.07
Asia	88.66	4.76	6.58
World	67.14	17.23	15.62

Source: International Tobacco Growers' Association (1995), *Tobacco & The Environment*, Tobacco Briefing, April, East Grinstead, p.2.

Flue-cured tobacco   As given in Table 9.2, among the sources of energy used to produce heat in flue-cured tobacco, coal is the most important source, while 'oil burners are in general use around the world', and 'wood is a very widely used fuel' (Akehurst, 1981). Related to the flue produce in 1993, coal accounts for about two thirds, while oil/gas and wood have about equal shares in the remaining third. The differences between the continents are considerable: while the use of wood is of minor importance in Europe and northern hemisphere America, it is widely used in South America, Africa and – to a lesser degree – in Asia.

As compared to the breakdown for 1993 (15.6%), a higher value of wood usage (18.8%) was found if related to the annual average production in 1991 to 1995. From Table 9.3, it could be seen that this is mainly due to a larger share of African flue using wood. In total, wood was found to be widespread among two thirds out of a total of 82 flue producing countries.[4]

Table 9.3 Dimensions of wood-based curing of flue, 1991/95

Region	Fuelwood for curing as % of flue tobacco cured	No. of flue producing countries all	using wood
North & Central America	1.88	12	9
South America	95.18	10	10
Europe (incl. Turkey)		16	
Africa	44.41	20	19
Asia	10.07	24	18
World	18.84	82	56

Sources: Adapted from Table 6.4; United States Department of Agriculture, Foreign Agricultural Service (1994, 1997), *Tobacco Circular Series*, FT-9407, FT-9707, USDA/FAS, Washington.

Table 9.4 provides a modal split of the sources of energy per country. Only about one third of the growing countries were found to make exclusive usage of firewood in curing flue (all of them developing nations), while another third has combined uses of wood and other sources of energy (such as coal in Zimbabwe and Argentina), and the remaining third makes exclusive usage of other energy sources than wood (most of them developed countries, but also China with usage of bituminous coal).

In summary, most tobaccos (i.e. more than 60% in 1991/95) are flue-cured using heat from different sources of energy. From the average annual world production of flue in 1991/995 (i.e. 4.6 million tons), about 12% was found using wood as a curing fuel as compared to around 10% in 1993 (ITGA, 1997). Wood is widely used in growing countries of the developing world. ITGA (1997) claims that 'in many tobacco-growing countries, wood

Table 9.4 Individual country sources of energy for curing flue

Number of growing countries	100%	Fuelwood as % of flue tobacco cured				
		90%	67%	50%	36%	30%
Africa	18[a]					1[b]
America		16[e]	1[d]			
Asia & Oceania	2[c]	7[f]			9[g]	2[h]
World (N=56)	27	16	1	9	2	1

[a] Angola, Benin, Congo/Zaire, Ethiopia, Ghana, Kenya, Madagascar, Malawi, Mali, Morocco, Mauritius, Mozambique, Nigeria, Reunion, Sierra Leone, Tanzania, Uganda and Zambia
[b] Zimbabwe [c] Brazil, Honduras [d] Argentina
[e] Chile, Colombia, Costa Rica, Dominican Republic, Ecuador, El Salvador, Guatemala, Guyana, Haiti, Jamaica, Mexico, Nicaragua, Peru, Trinidad & Tobago, Uruguay and Venezuela
[f] Pakistan, India, Myanmar, Cambodia, Laos, Malaysia and Philippines
[g] Cyprus, Iran, Jordan, Korea/N., Korea/S., Syria, Thailand, Viet Nam and Yemen
[h] Bangladesh, Sri Lanka

Sources: Akehurst, B.C. (1981), *Tobacco*, Longman, London, p. 250; Baguley, M. and Inglis, C. (1990), *Report on fuelwood consumption and plantation establishment survey for Tabacalera Hondurena*, LTS International Ltd., Edinburgh, p.8; *ibid*. (1990), *Report on fuelwood consumption and plantation establishment survey for Nigerian Tobacco Company Ltd.*, LTS, Edinburgh, p.6; Fraser, A.I. (1986), *The use of wood by the tobacco industry in Argentina*, International Forest Science Consultancy, Edinburgh, p.11; *ibid*. (1986), *The use of wood by the tobacco industry in Brazil*, Edinburgh, p.13; Fraser, A.I. and Bowles, R.C.D. (1986), *The use of wood by the tobacco industry in Kenya*, IFSC, Edinburgh, p.7; *ibid*. (1986), *The use of wood by the tobacco industry in Zimbabwe*, IFSC, Edinburgh, p.5; Gossage, S.J. (1997), *Land use on the tobacco estates of Malawi*, ELUS, Lilongwe, 1997, p.29; Inglis, C. and MacDonald, C. (1991), *Report on fuelwood consumption and plantation establishment survey for the Bangladesh Tobacco Company*, LTS, Edinburgh, p.6; International Tobacco Growers' Association, *Tobacco & The Environment*, Tobacco Briefing, ITGA, East Grinstead, p.2; ibd. (1997), *The use of woodfuel for curing tobacco*, Tobacco Growers Issues Papers, No. 11, ITGA, East Grinstead, p.3.

fuel is being used less', but as a renewable resource and where alternatives are not easily available or expensive, 'it will continue to play an important role in tobacco-growing countries for decades to come'. It is assumed that the expansion of other sources than wood 'will largely come in countries which use these fuels as primary sources of energy for flue-curing already' (ITGA, 1997).

Poor efficiency of fuelwood used

While up to the mid-19th century, tobaccos had been naturally cured, wood was the obvious source of flue curing in the early days since the first application (or discovery) of the curing technique in 1852. The standard curing facilities (tobacco barns or kilns) were described by Akehurst (1981) to have retained 'a remarkable degree of uniformity' around the world. He stresses that the fundamental principles, i.e. producing heated air to dry up the leaves, 'have changed not at all since the introduction of metal flues about the time of the American Civil War', and that 'the majority of the world's flue-cured tobacco is still cured in the conventional tall barn'.

According to tobacco industry commissioned reports, woodfuel is used with greater efficiency due to improved insulation and new furnace designs, and tobacco growers are leading the way on major improvements in the use of agricultural wastes, i.e., compounding of crop waste in pellets and briquettes as fuel sources (ITGA, 1997). However, despite the current impetus given to the need for saving energy, Akehurst (1981) insists that the unchanged fundamentals, i.e., evaporating about 85 parts out of every 100 parts of field grown material, mean that even 'equipment of increasing sophistication ... is still consuming a very large amount of energy (...) which so far only allowed poor output per unit of energy'. It is estimated that the wastage of energy *en route* of the whole curing process ranges between 70% and 90% of all energy input. While several approaches towards reducing energy (and cost requirements) are in investigation, 'most are concerned only with modification of existing principles in the hope of maintaining the basic *status quo* a little longer' (Akehurst, 1981).

The conventional and widely used convection type of barn makes use of natural air circulation, while so-called forced draught systems use heat exchangers driven by artificially produced energy with lesser fuel costs per unit of produce than with conventional barns. However, with regard to installation in rural areas of the developing world, it is important to note that 'electric power is necessary for all forced-draught systems and to operate automatic fuelling devices which use coal' (Akehurst, 1981).

In summary, unless the prevailing tobacco definition standards for flue were changed drastically, it could be expected that furnaces of the conventional convection type of barn will have to be changed to a design which uses compounds of agricultural waste instead of alternative fuels (such as coal, oil, gas, electricity, infrared or solar energy). The change in the economics of fuel choice, and hitherto the future use of wood in the developing world, will to a large extent be regulated by the true (and increasing) costs of cutting and transport.

Self-sufficiency in wood and the commons issue

The crucial point was laid down by Fraser (1986) that, as a forest product, wood is commonly taken from natural forests and woodlands and regarded as 'free good' in the developing world. This means that no payment or contribution is made towards the cost of replacement (even when sold in the free market, in general the price of wood only reflects the harvesting cost). Only where shortages have developed, the market price is rising to a level where investment in plantation forests is becoming attractive. However, 'there seems little hope that such investment will take place on a sufficient scale to meet future fuelwood demand until the most of the natural forest has been destroyed' (Fraser, 1986).

With varying degrees of success, some governments and tobacco buyers are already active in promoting efforts of afforestation (or reforestation), partly by negotiating leaf-growing contracts which are conditional upon farmers planting fuelwood, partly by integrating tobacco wood lots into programmes of social forestry (ITGA, 1997). As a rule of thumb, the following ratio is generally held to be acceptable now and for the future: one hectare of sustainably managed tree plantations provides sufficient energy to cure one hectare of (flue) tobacco (ITGA, 1996a).

However, despite scarcity of data, there is an indication from Table 9.5 that only less than half of all flue tobacco growers have woodfuel plantings on their own. The percentage of flue farmers disposing of own wood supplies (and assumedly being self-sufficient in wood) ranges between 5-7% (such as in Poland or Tanzania) and 100% (such as in Kenya and Congo/Zaire). If one is to expect that full self-sufficiency in wood exists if own wood supplies are given, about 60% of flue tobacco farmers will have to rely upon other than own sources of wood, i.e., assumedly taking wood from open accessible ('common') natural forests and woodlands.

The point was made by Chapman (1994) that issues of land tenure – 'so variable and insecure throughout much of the Third World' – might have implications for projects of reforestation (and afforestation), 'for only people with secure title to land are likely to consider making the long-term

Table 9.5   Percentage of flue growers who have woodfuel plantings

Argentina*	10%	Sierra Leone	25%
Brazil	82%	Sri Lanka	75%
Congo/Zaire	100%	Tanzania	7%
Honduras*	10%	Uganda	80%
Kenya	100%	Zambia	10%
Malawi	20%	Zimbabwe	70%
Nigeria	30%		
Pakistan	10%		
Poland	5%	Total (world)	42%

* Estimation values

Sources: Baguley, M. and Inglis, C. (1990), Report on fuelwood consumption and plantation establishment survey for Tabacalera Hondurena, LTS International Ltd., Edinburgh, p. 8,16; ibd. (1991), Report on fuelwood consumption and plantation establishment survey for Nobleza-Picardo S.A.I.C.y F, Argentina, LTS International Ltd., Edinburgh, p. 10; International Tobacco Growers' Association (1997), The use of woodfuel for curing tobacco, Tobacco Growers Issues Papers, No. 11, ITGA, East Grinstead, p.6.

investment required for tree growing'. In understanding why the manner of use of common property resources (CPRs) is important and why they are particularly vulnerable to induced degradation, Blaikie and Brookfield (1994) noted that 'where CPRs are encroached upon and privatised through enclosure, the remaining areas have to carry the added displaced load of the CPR users'. A growing body of studies on social and environmental change in middle mountain and highland areas of the developing word (i.e., the natural environments where tobacco is commonly grown) suggest increasing vulnerability of regional populations as well as land degradation that may accompany the transformation from traditional common property arrangements to privatisation and commercialisation (Turner, *et al.*, 1995). Using this notion of pronounced land degradation during periods of land tenure transformation, it was shown by Geist (1998c) in the case of a highland tobacco area with the crop grown under varying tenurial regimes that large usage of wood (and related land degradation) could be attributed to the growing of flue under next to private leasehold arrangements (estates') holding more than half of all cultivable land previously under customary right.

Usage of polewood

For all tobaccos grown, barn facilities are required for the adequate curing, mid-rib drying and storage of tobacco such as curing barns, drying sheds and holding barns. The facility requirements also include loading sticks and tier poles (inside the barn) and raised above the ground platforms made out of poles or planks on which, after curing and conditioning the leaves, bulking takes place to allow for air circulation.

Flue-cured tobacco At least one curing barn (5m x 5m x 12m) is required per hectare of flue-cured tobacco. According to ITGA (1995), 'more than four-fifths of the world's flue-cured tobacco is dried in conventional barns constructed from a variety of materials', i.e. logs of wood with mud filling, wooden-frames, cement blocks, bush poles and mud. Akehurst (1981) notes that 'wood is the main material in the USA and Canada, often in the case of frame buildings', while 'in Zimbabwe, the barns are almost all of burnt brick'. There is a general tendency towards more simple constructions as the scale of production gets smaller. It had also been noted, that 'the standard tobacco barn (...) has retained a remarkable degree of uniformity', and though the requirement of an airtight building is essential, 'in practice very many conventional barns are not airtight' (Akehurst, 1981). Inside the conventional barn are the typical layers of tiers or racks spaced according to local stick length. Commonly, the foundation is laid in concrete, and furnace openings, ventilators and doors should be covered by concrete lintels for strength and for ease of maintenance and repairs. Once the curing is completed, the barn is emptied and tobacco is bulked or rough baled on platforms of poles or planks.

Fire-cured tobacco Seven standard curing barns of 4m x 2-3m x 2-3m with air tight walls and two fire pits as well as two large holding barns for mid-rib drying are required to cure leaf from about one hectare as in the case of Malawi (Mittawa, 1985; Kamvazina, 1994). According to Akehurst (1981), 'barns in the USA are of tight timber-frame and board construction', while 'barns in Africa are usually very simple mud or mud-and-pole structures'. After curing, the conditioned leaf should not be left hanging in the barn, but removed from the sticks and bulked.

Air/sun-cured & oriental tobaccos As Akehurst (1981) noted, 'it is now very common practice to wilt strung tobacco in the shade ... before exposure to the sun'. A minimum of four standard wilting and yellowing barns (55 m long) are required per 0.4 hectare of tobacco as well as an open-sided curing shed of 25m x 3m x 3m per half a hectare (for the purpose of air-curing). The mechanics of sun-curing are many and varied. They range from laying the strings on the ground to sophisticated movable frames (which are wheeled in and out of permanently roofed shelters) and to simple open air curing racks (or closable curing tents). In addition to curing tents, holding barns are required to store the cured leaf such as in Malawi (Mittawa, 1985; Kamvazina, 1994). After curing, the leaf may be bulked or hung in garlands from a roof.

Burley & light air-cured tobaccos   The curing is 'traditionally (done) in large wooden buildings with provision for extensive side-wall ventilation (...) In countries where the materials are readily available, adequate structures can be made from bush poles and thatch' (Akehurst, 1981). Both curing sheds and holding barns (drying sheds) are required. While one curing shed of 33m x 6m x 5m will provide enough curing accommodation for about one hectare of burley such as in Malawi (Mittawa, 1985), there is some variation of the definite barn requirement, ranging from 40 to 250 metres in length for one hectare of burley (TRIM, 1994). It had been found that the mean number of sheds required on a large scale burley farm ('estate') in southern Malawi was 35, i.e., requiring between about 310 and next to 148,000 poles (Geist, 1997b). Once the leaf is cured, it should not be left in the curing shed any longer, but transferred to large drying sheds whilst the mid-rib dries out. While the common curing structure (such as in Malawi) is a low barn built out of indigenous timber poles and thatch, holding barns are normally large buildings constructed of (eucalyptus) plantation tree poles and corrugated iron roofing (used for both drying mid-ribs and storage of leaf transferred from the curing barns after the lamina has cured and dried). Since a load of freshly cut tobacco is very heavy, the construction standards must be high. Taking again the example of Malawi, it has been found that most of the curing barns 'are sub-standard in terms of materials and construction technique. As a result, most growers have had to undertake frequent maintenance work on existing structures or reconstruct barns annually', while this problem 'contributes to high demand for construction poles' (TRIM, 1994).

Dark air-cured tobacco   Any shelter, mostly sophisticated wrapper curing sheds, will protect the leaf from sun, rain and direct exposure to air. For example, a typical Sumatra structure is made of rough timber covered by well-laid palm leaves completed by a suitable arrangement of tier poles across which to suspend the sticks of tobacco. Akehurst (1981) noted that 'Connecticut sheds are of wood (with) Canadian sheds (being) similar to those in the USA'. Most of the growing countries in the developing world have sheds of varying degrees of simplicity which are often made of local materials, while 'European dark tobaccos are cured in the well-built structures which the climate demands' (Akehurst, 1981).

In addition to the fuelwood and polewood usage by tobacco farming, it was suggested by Gossage (1997) to further include the use of wood for domestic purposes by resident populations on large commercial farms or 'estates' (such as in Malawi), i.e., the private wood consumption by tenants, temporary or permanent workers, employees and their families. However, these case are not considered here.

Empirical rates of wood usage[5]

Fraser (1986) was first to create and apply global 'index units of wood consumed per ton of tobacco' on the basis of data drawn from seven growing countries (i.e., Argentina, Brazil, Kenya, Malawi, Zimbabwe, India and Thailand), while extrapolations were made to 69 major developing nations growing tobacco. The values given (e.g., 6.4 m^3/ton for the firewood usage of flue) differed considerably from other values applied during those times. For example, Goodland, et al., (1984) used a mean value of 55 m^3/ton of flue, however explaining that 'there is considerable regional variation ... (and) estimates vary from 70 m^3/ton in Tanzania to 20 m^3/ton in the Philippines'.

Following a 'ten year drought of information on the issue of deforestation caused by tobacco growing and curing in developing countries' (Chapman, 1994), three case reports on the environmental (as well as social) consequences of tobacco farming in Kenya, Tanzania and Uganda were published by Waluye (1994), Kweyuh (1994) and Muwanga-Bayego (1994). Again, the values specified ranged far above the ones given by Fraser (1986). As a major default, none of the ITGA studies on curing, wood usage and deforestation has ever provided updated figures comparable to the ones specified before. The only and last usage rate was provided by ITGA (1997) as a 'median' value of fuelwood consumption instead of the mean, minimum or maximum values used before. For this reason, a compilation of (comparable) wood usage data will be attained in the following.

Fire-cured tobacco   Only two mean values had ever been specified in the case of fuelwood consumed by (dark) fire-cured tobacco, i.e., Fraser (1986) for the 1983/84 growing season and Geist (1997a,b) for the 1995/96 season, all of them taken from African growing countries. As could be seen from Table 9.6, the values differ considerably, i.e., 8 versus 38 stacked cubic metres per ton of (cured) produce. However, since no other data were available, the mean of both is taken to approach the (assumed) mean fuelwood usage of fire-cured tobacco as 23 stacked cubic metres per ton of produce world-wide (i.e., 8 plus 38 divided by 2).

Table 9.6   Rates of fuelwood usage by fire-cured tobacco

Malawi (1983/84)	3.7 kg/kg	or	9 m^3/ton
Kenya (1983/84)	3.0 kg/kg	or	7 m^3/ton
Malawi (1995/96)	16.2 kg/kg	or	38 m^3/ton
Tanzania (1995/96)	16.0 kg/kg	or	37 m^3/ton
Total (1991/95)	9.7 kg/kg	or	23 m^3/ton

Sources: Fraser, A.I. (1986), The use of wood by the tobacco industry and the ecological implications, International Forest Science Consultancy, Edinburgh, p.17; Geist, H. (1997a), Tobacco growers of Songea District and their miombo environment (Ruvuma Region, Southern Tanzania): Preliminary results, Institute of Geography, University of Düsseldorf, p.16; Geist, H. (1997b), Tobacco growers of Namwera RDP/Malawi: A study of social and environmental change, Analytical draft report, Institute of Geography, University of Düsseldorf, p.35.

The index unit of polewood consumed by fire-cured for 1983/84 was 3.4 m^3/ton as specified by Fraser (1986). In this estimation, a recent rate of polewood usage is assumed to be only around 0.5 (stacked) cubic metres per ton, i.e., about five times the polewood usage of flue as specified by Fraser (1986).

Flue-cured tobacco   The data as given in Table 9.7 suggest a marked downward trend in the consumption of firewood by African flue-cured tobacco since about 20 to 30 years. The mean consumption of stackwood per ton of produce in the first half of the 1990s amounts to 28 m^3/ton (i.e., 38 plus 17 divided by 2), which is only half of what it had been in the 1970s and 1980s, i.e., 59 m^3/ton (25 plus 92 divided by 2). It should be mentioned here that only such consumption data were considered that could either be drawn from field reports of government agencies (and related bodies) or from scientific literature with a stated research design. Thus, data were excluded or not considered that originate from a more or less casual interviewing of farmers as done by Kweyuh (1994), Muwanga-Bayego (1994) and Waluye (1994).

Table 9.7 Rates of fuelwood usage by flue-cured tobacco in Africa

Tanzania (ca. 1970)				148 m³/ton
Tanzania (ca. 1970)				95 m³/ton
Tanzania (1970s)				122 m³/ton
Tanzania (1970s)				77 m³/ton
Tanzania (mid-1970s)				83 m³/ton
Tanzania (mid-1970s)				50 m³/ton
Tanzania (late 1970s)				70 m³/ton
Malawi (1983/84)	13.0 kg/kg	or	30 m³/ton	
Zimbabwe (1983/84)	10.8 kg/kg	or	25 m³/ton	
Kenya (1983/84)	8.0 kg/kg	or	19 m³/ton	
Total	10.6 kg/kg	or	25 m³/ton	92 m³/ton
Tanzania (1989/90)	14.2 kg/kg	or	33 m³/ton	
Nigeria (1990)	26.8 kg/kg	or	62 m³/ton	
Malawi (ca. 1990)				16 m³/ton
Malawi (1995/96)				17 m³/ton
Malawi (1995/96)	8.0 kg/kg	or	19 m³/ton	
Total	16.3 kg/kg	or	38 m³/ton	17 mt³/ton

Sources: Baguely, M. and Inglis, C. (1990), Report on fuelwood consumption and plantation establishment survey for Nigerian Tobacco Company Ltd., LTS, Edinburgh, p.3; Bernard, M.P. (1990), Preliminary economic considerations of advanced curing systems in Malawi, Lilongwe, p.1; Boesen, J. and Mohele, A.T. (1979), The 'success story' of peasant tobacco production in Tanzania, CDR Publications No. 2, Uppsala, p.92; Chuffney (1979), cited by Kanshahu, A. (1980), Smallholder tobacco production improvement project, Dar Es Salaam, A-9; Fraser, A.I. (1986), The use of wood by the tobacco industry in Zimbabwe, IFSC, Edinburgh, p.5; ibd. and Bowles, R.C.S. (1986), The use of wood by the tobacco industry in Kenya, IFSC, Edinburgh, p.7; ibd. (1986), The use of wood by the tobacco industry in Malawi, IFSC, Edinburgh, p.13; Goodland, R.J.A., Watson, C. and Ledec, G. (1984), Environmental management in tropical agriculture, Westview Press, Boulder, p.58; Geist, H. (1997b), Tobacco growers of Namwera RDP/Malawi, Düsseldorf, p.17; Gossage, S.J. (1997), Land use on the tobacco estates of Malawi, ELUS, Lilongwe, p.30; Siddiqui, K.M. and Rajabu, H. (1996), 'Energy efficiency in current tobacco-curing practice in Tanzania and its consequences', Energy, vol. 21(2), p.143; Temu, A.B. (1979), Fuelwood scarcity and other problems associated with tobacco production in Tabora region, Faculty of Forestry Record No. 12, Morogoro, p.13; World Bank (1977) cited by Kanshahu, A. (1980), op. cit.

In contrast to the situation in Africa, generally lower levels of firewood consumption by flue-cured tobacco were found in Asian and Latin American growing countries. Over the decades, they seem to have more or less remained the same.

As it is shown in Table 9.8, the average consumption of stackwood per ton of flue produce in Asia is assumed to amount to about 19 stacked cubic metres, i.e., the mean of the value of the 1970s and 1980s (20 m^3/ton) and the first half of the 1990s (17 m^3/ton) taken together (i.e., 20 and 17 divided by 2). Asia is the single case where, due to data scarcity, the (recent) mean usage rate is constructed from values of all periods of time.

Table 9.8   Rates of fuelwood usage by flue-cured tobacco in Asia

Philippines (late 1970s)			20 m^3/ton	
India (1983/84)	5.3 kg/kg	or	12 m^3/ton	
Thailand (1983/84)	11.4 kg/kg	or	27 m^3/ton	
Total	11.0 kg/kg	or	20 m^3/ton	20 m^3/ton
Bangladesh (1990)	7.2 kg/kg	or	17 m^3/ton	
(Total)	(7.2 kg/kg)	or	(17 m^3/ton)	

Sources: Fraser, A.I. (1986), *The use of wood by the tobacco industry and the ecological implications*, IFSC, Edinburgh, p.17; Goodland, R.J.A., Watson, C. and Ledec, G. (1984), *Environmental management in tropical agriculture*, Westview Press, Boulder, p.58; Inglis, C. and MacDonald, C. (1990), *Report on fuelwood consumption and plantation establishment survey for the Bangladesh Tobacco Company*, LTS, Edinburgh, p.4.

According to the date in Table 9.9, the mean usage of wood in Latin America has estimatedly been on a slight increase from 13 to 15 m^3/ton. Thus, if the three continental averages were added, the global mean amount of fuelwood to produce one ton of flue-cured tobacco in the 1990s is around 21 stacked cubic metres, i.e., 28 (in Africa) plus 19 (in Asia) plus 15 (in Latin America) divided by three. Again, the value-derived ranges above the one provided by Fraser (1986), i.e., 6.4 m^3/ton.

In order to assess the polewood usage of flue, an overall rate of 0.1 stacked cubic metres per ton of produce was applied (except for Zimbabwe and Malawi where brick built barns are reported). This value goes hand in hand with the one specified by Fraser (1986), i.e. 0.14 m^3/ton.

Table 9.9   Rates of fuelwood usage by flue-cured tobacco in Latin America

Brazil (1983/84)	5.9 kg/kg	or	14 m³/ton
Argentina (1983/84)	4.8 kg/kg	or	11 m³/ton
Total	5.4 kg/kg	or	13 m³/ton
Honduras (1990)	7.8 kg/kg	or	18 m³/ton
Argentina (1990)	5.2 kg/kg	or	12 m³/ton
Total	6.5 kg/kg	or	15 m³/ton

Sources: Fraser, A.I. (1986), The use of wood by the tobacco industry in Brazil, IFSC, Edinburgh, p.7; Fraser, A.I. (1986), The use of wood by the tobacco industry in Argentina, IFSC, Edinburgh, p.5; Baguely, M. and Inglis, C. (1990), Report on fuelwood consumption and plantation establishment survey for Tabacalera Hondurena, LTS, Edinburgh, p.5; Baguely, M. and Inglis, C. (1990), Report on fuelwood consumption and plantation establishment survey of Argentina, LTS, Edinburgh, p.4.

Other tobaccos   Taking the polewood usage of burley tobacco as representative for other naturally cured varieties grown, Gossage (1997) gave a range of values for the mid-1990s, i.e., between 2.5 stacked cubic metres per ton in the case of technologically improved burley barns and 5.0 m³/ton in the case of ordinary or traditionally built burley barns. Since the national share of technologically [un]improved barns will be hard to quantify on a global scale, a preliminary breakdown as given in Table 9.10 is attained to allow for differences of climate and technological standards on a country level, e.g., well-built structures under climates demanding such.

Table 9.10   Rates of polewood usage in naturally cured tobaccos

	Country status of development (income)		
m³/ton of tobacco	high	middle	low
Tropical countries	2.5	3.8*	5.0
Non-tropical countries	0.5*	1.0*	2.5

* Estimation values

Source: Adapted from Gossage, S.J. (1997), *Land use on the tobacco estates of Malawi*, Estate Land Utilisation Survey, Lilongwe, p.30.

Thus, national estimation values of polewood usage estimatedly range between 0.5 m³/ton in high income countries of the non-tropical zone and 5.0 m³/ton in low income countries of the tropical zone. While the value given by Fraser (1986) for burley tobacco, i.e., 3.01 m³/ton, fits the range constructed here for most of the developing nations, the overall value of 0.07 m³/ton for oriental, dark air/sun-cured, light air-cured and dark air-cured (cigar) tobacco is well below the ones specified.

Specific wood consumption on a global scale

In a further step, the national values of wood usage in the 1990s were applied to the national annual mean amount of tobaccos produced from 1991 to 1995 (USDA/FAS, 1994; 1997). In cases where no national values were available, the respective continental means were taken.

Table 9.11 summarises the consumption of wood in tobacco farming. Thus, around 20 stacked cubic metres are used to produce one ton of tobacco as a mean global average for the 5-years period 1991-95. The combined uses of firewood and polewood by artificially cured varieties, i.e., flue and fire-cured tobaccos, by far outweigh wood consumption of naturally cured tobaccos. Due to country-specific variations of the sources of energy used in curing and due to varying mixtures of construction materials used in barn construction, the usage of wood will range between 0.5 and about 62 stacked cubic metres per ton of (cured) produce, while the most frequent value (mode) is just one.

As for comparison, and to give a rough indication of the order of magnitude involved, Table 9.12 specifies values of fuelwood uses in other rural industries of the developing world. Thus, one cubic metre of wood provides heating and cooking for one person for a year, smokes one ton of fish, burns 3,000 bricks, or cures 50 kg of tobacco according to this estimation.

The mean of around 20 stacked cubic metres per ton of produce translates into a (mean) solid wood measure of 9 kilograms of wood per 1 kilogram of (cured) tobacco ranging between 5 and 12 kg/kg.[5] In a previous study which until recently had been regarded to be 'the definitive study on the use of wood by the tobacco industry' (ITGA, 1995), the average value from a sample of seven growing countries was found to be around 8 kg wood per kilogram of tobacco, ranging between 5 and 13 kg/kg for individual country averages in 1983/84 (Fraser, 1986). As a comparison, though different sample sizes were used, since about one decade the specific wood consumption could have more or less remained the same, if not showing a slight tendency towards increase (while this could also be attributable to the different sample sizes used).

Table 9.11   Annual wood consumption in tobacco farming, 1991/95 (stacked cubic metres per ton of cured tobacco)

	Mean	(range)	Median	Mode
Artificially cured tobaccos				
Flue-cured, total	21.3	(12.1-62.1)	19.1	*
...fuel	21.2		19.0	15.0
...poles	0.1		0.1	0.1
Fire-cured, total	24.3	(23.5-37.5)	23.5	23.5
...fuel	23.8		23.0	23.0
...poles	0.5		0.5	0.5
Naturally cured tobaccos				
Burley	3.2	(0.5-5.0)	3.8	3.8
Oriental	2.6	(0.5-5.0)	2.5	1.0
Dark air/sun-cured	3.5	(0.5-5.0)	3.8	5.0
Dark air-cured	3.4	(0.5-5.0)	3.8	3.8
Light air-cured	3.8	(0.5-5.0)	3.8	3.0
All tobaccos	19.9	(0.5-62.1)	18.9	1.0

* Multiple modes

Sources: Adapted from Tables 6 to 10 using USDA/FAS production figures; United States Department of Agriculture, Foreign Agricultural Service (1994, 1997), *Tobacco Circular Series*, FT-9407, FT-9707, USDA/FAS, Washington.

In contrast to this, tobacco industry commissioned reports claim that during the same period of time 'there have been dramatic reductions in the consumption of woodfuel' (ITGA, 1997), and that 'through the establishment of renewable, energy-efficient and regularly harvestable sources in managed tree plantations, (developing countries) will, however, stabilise the impact they make on the deforestation question' (ITGA, 1996a). On the basis of a recent survey, undertaken in 1996 among 'tobacco growers' organisations or related agri-businesses', the 'median ratio of woodfuel used for flue-curing was found to be 5.5 kg wood: 1 kg of cured tobacco' with the sample coming from 14 producer countries (ITGA, 1997).

Table 9.12 Fuelwood uses in rural industries of the developing world

One (stacked) cubic metre of wood ...
- * provides heating and cooking for one person for a year
- * brews 400 litres of beer
- * smokes one ton of fish
- * cures 50 kilogram of tobacco
- * produces 3 kilogram of salt[a]
- * burns 3,000 bricks[b]

a) The amount produced by Tanzanian salt-makers in half a day.
b) One third of the bricks needed to build a standard rural house.

Source: Booth, A., and Clarke, J. (1994), 'Woodlands and forests', in M. Chenje and P. Johnson (eds.), *State of the environment in Southern Africa*, Southern African Research & Documentation Centre, Harare, Masero.

However, from the data specified by ITGA (1997), the claims of a drastic reduction in wood usage (and tobacco-related environmental implications) have to be challenged for several reasons. First, the comparative value of Fraser (1986) for the 1983/94 growing season was specified as a mean (plus range), while ITGA's (1997) most recent figure is specified as a 'median' (and not mean) value, while neither a range nor individual country averages are given. Secondly, the figure of Fraser (1986) comprises wood usage of all artificially cured varieties, i.e. flue and fire-cured tobaccos, while the ITGA report merely considers flue-cured tobacco of assumedly large farms ('agro-businesses') with sophisticated technology and high fuelwood efficiency. Even if the median value of flue's specific wood consumption in 1996 is taken (5.5 kg/kg), the difference to the situation in 1991/95 as found here amounts to 50 percent as could be seen from the median value of flue as given in Table 9.11, i.e., 8.2 kg/kg (19.1 stacked cubic metres x 0.43).[5]

Wood consumption on a global scale

In a last step, the national values of specific wood consumption were first combined with the national sources of energy used, and later with the national degrees of self-sufficiency in wood. From Table 9.13, it could be seen that slightly more than half of the average annual global tobacco produced in about 120 growing countries in 1991 to 1995, i.e., 3.8 out of 7.5 million tons, makes usage of wood. Naturally cured tobaccos seem to outweigh artificially cured tobaccos (with combined polewood and firewood requirements) by factor 3, with dark air/sun-cured, flue-cured, burley and oriental tobaccos holding the largest single shares. The usage of wood was found to be prevalent among 116 out of a total of 118 growing countries with reported and variety specific tobacco production (USDA/FAS, 1994; 1997).

Table 9.13 Global annual tobacco production using wood in 1991/95 (in '000 metric tons of dry weight)

	World production		... using wood	
Artificially cured tobaccos	4,681	(61.8%)	949	(12.6%)
Flue-cured tobacco	4,627	(61.1%)	895	(11.9%)
Fire-cured tobacco	54	(0.7%)	54	(0.7%)
Naturally cured tobaccos	2,829	(37.7%)	2,829	(37.6%)
Burley	893	(11.9%)	893	(11.9%)
Dark air/sun-cured	1,094	(14.6%)	1,094	(14.6%)
Oriental	651	(8.7%)	651	(8.7%)
Dark air-cured	114	(1.5%)	114	(1.4%)
Light air-cured	77	(1.0%)	77	(1.0%)
All tobaccos	7,510	(99.5%)*	3,778	(50.2%)

* The sum may not end up to 100 percent due to rounded production figures

Sources: Adapted from Tables 6 to 10; USDA/FAS production figures; United States Department of Agriculture, Foreign Agricultural Service (1994, 1997), *Tobacco Circular Series*, FT-9407, FT-9707, USDA/FAS, Washington

In Table 9.14, the usage of wood by 3.8 million tons of tobacco is translated into solid measures amounting to a total of 11.4 million tons of solid wood consumed each year in 1991 to 1995. On the basis of varying specific rates of wood consumption, the wood requirement of artificially cured tobaccos now outweighs that of naturally cured tobaccos by factor 2, with flue-cured tobacco holding the largest single (and uniquely high) share of around 60%. Accounting for degrees of self-sufficiency in wood, half of the solid wood requirement is estimatedly taken from other than own supplies, with fuelwood requirements by flue tobacco holding the leading and largest single share.

Table 9.14 Global annual wood requirements of tobacco in 1991/95 (in '000 metric tons of solid wood)

	Total solid wood		...not taken from own supplies	
Artificially cured tobaccos	7,420	(64.8%)	3,401	(59.6%)
Flue-cured tobacco	6,849	(59.8%)	3,028	(53.1%)
... fuel	6,810	(59.5%)	3,011	(52.8%)
... poles	38	(0.3%)	17	(0.3%)
Fire-cured tobacco	571	(5.0%)	373	(6.5%)
... fuel	560	(4.9%)	365	(6.4%)
... poles	12	(0.1%)	7	(0.1%)
Naturally cured tobaccos*	4,018	(35.1%)	2,298	(40.4%)
Burley	987	(8.6%)	575	(10.1%)
Dark air/sun-cured	2,030	(17.7%)	1,150	(20.2%)
Oriental	696	(6.1%)	404	(7.1%)
Dark air-cured	181	(1.6%)	102	(1.8%)
Light air-cured	124	(1.1%)	67	(1.2%)
All tobaccos	11,437	(99.9%)	5,698	(100.0%)

* only polewood

Sources: Adapted from Tables 4, 5 and 13 using USDA/FAS production figures; United States Department of Agriculture, Foreign Agricultural Service (1994, 1997), *Tobacco Circular Series*, FT-9407, FT-9707, USDA/FAS, Washington

Thus, and in summary, though flue-cured tobacco which is the main ingredient of (light) blended cigarettes makes up only 12% of the global tobacco using wood, it accounts for more than half of all wood consumed (and around half of the wood assumedly taken from natural environments).

As for comparison, and to give a rough indication of the order of magnitude involved, the total amount of wood required annually in tobacco farming during the first half of the 1990s, i.e., 11.4 million tons of solid wood (or the equivalent of 26.6 million stacked cubic metres), constitutes the equivalent of burnt bricks to build nearly 9 million standard rural houses per year in the developing world. To put this into a global perspective, a comparison between the estimated wood consumption of tobacco farming and the total recorded wood consumption shows that the tobacco sector is a relatively minor consumer, i.e., only accounting for less than 1 percent of the wood consumed directly (Fraser, 1986; FAO, 1997a).

Environmental implications

On a global scale, there is reason to assume that tobacco growing has an *indirect* impact by expanding the global agriculture frontier more rapidly than other crops do. While, on a national and regional scale (and even in major geographical areas), tobacco farming well exerts a *direct* impact being the leading commercial user of wood (however, still lacking behind domestic consumption) so that the notion of wood-use related deforestation could fully be applied. In both cases, tobacco's significance for land-use and land-cover changes is evident.

Causes of tobacco-induced deforestation

While so far only the usage of wood has been considered, several causes of the removal of forest and woodland cover by tobacco farming in the developing world could be isolated and identified.

First, tobacco growing is part of the overall expansion of agricultural land which partly, if not mostly, occurs at the expense of vegetation cover. Not only in historical terms, tobacco constitutes a 'pioneer crop' (Goodman, 1995) at the agriculture/forest frontier, but also with regard to the recent colonisation of virgin land in remote areas of the developing world. As for example, ITGA (1997) emphasises in the case of newly induced tobacco expansion in Zambia that the crop 'has proved the most effective means of opening up lands for development. It is frequently the first crop on these lands'. This could partly be explained by tobacco's high value to generate easy cash income ('brown gold').

Secondly, the soil-depleting nature of the crop could be interpreted as having a special impetus towards the clearance of new land in order to provide fresh and fertile soils. In the past, and partly even today, this is done by means of deforestation (Goodland, *et al.*, 1984; Chapman, 1994).

Thirdly, in order to create a smoke product reaching out for taste, aroma, combustibility and high levels of nicotine, the usage of wood for curing the harvested green leaves plays a major and essential role in exerting pressures upon natural resources, particularly in rural growing areas where no alternative and cost-effective energy sources to wood exist. By having formatted the annual usage of wood for the curing of tobacco in

1991 to 1995, and to contribute to the 'great debate and uncertainty' (Williams, 1994a) of the world-wide felling of trees, the resulting consequences upon land-use/cover changes could indicatively be derived as follows.

Dimensions of wood use-induced deforestation

On the basis of wood consumption data as presented particularly in Tables 9.11, 9.13 and 9.14, it had been estimated by Geist (1999b) that about 200,000 hectares of (dry) natural forests and woodlands were eliminated each year during the 5-years period of 1991 to 1995. This type of recent (and not historical) deforestation was found to be prevalent in 66 growing countries nearly all of them situated in the developing world (i.e., in about half of all growing countries of the world).

In total, tobacco related deforestation amounts to around 2% of total net losses of forest cover as specified by FAO (1997) for the period considered, or to about 5% of total national deforestation in the growing countries affected. Environmental criticality such as inadequate forest cover to supply the respective population with fuelwood could be identified in 35 producer countries with an estimated 'serious', 'high' and 'medium' degree of tobacco related deforestation (Geist, 1999b).

Thus, the alarming observation by Fraser (1986) could largely be verified in 'that it is important to note that a high proportion of the tobacco growing areas in developing countries lie within parts of the world identified by FAO as being in wood deficit or prospective wood deficit situations'. Though surveying and extrapolating wood usage of tobacco from a rather small sample of growing countries, Fraser (1986) gave no formatted system to wood use related deforestation. However, by correlating tobacco production and forest availability, he concluded that the 'figures suggest that most Asian tobacco growing countries, and selected African countries have general fuelwood shortages and are therefore likely to experience accelerating deforestation'.

In updating these findings, a recent global assessment of tobacco related deforestation (Geist, 1999b) proves that not only in Africa (i.e., mostly in the southern part) and Asia (i.e., in the eastern, southern and middle eastern part), but also in South and Caribbean America felling of trees for tobacco and (assumedly) related degrees of environmental criticality could be identified. As a matter of fact, while FAO (1997b) data show that the present share of global tobacco in all arable land is relatively low (i.e., less than 1%), land under tobacco by far exceeds the global reference in major geographical areas of the developing world identified to be major tobacco growing regions such as East and Middle East Asia (1.5% each) or southern Africa (2.3%).

Thus, and in general, the indicative order of magnitude involved could well lead to qualify tobacco growing areas as 'regions at risk' (Turner, *et al.*, 1995). The hypothesis is put forward that tobacco is not only 'a major world crop' (ITGA, 1996b) and 'driving force for economic development' (Reemtsma, 1995), but also constitutes a major social and economic driver of land-use and land-cover changes particularly in dry forest and woodland ecosystems of the developing world.

Conclusions

The location of tobacco farming areas under semihumid to semiarid climates of the developing word, falling there more under highland than lowland ecosystems (with dryland as well as highland environments bearing striking features of ecological fragility), suggests far-reaching and mostly critical consequences for the sustainable management of man/land relations.

While highland areas are vulnerable ecosystems in that sustainable watershed management of forest cover is essential in order not to induce large-scale on-site effects (such as erosion) and off-site effects (such as flooding or climatic change), dryland areas are among the world's most fragile and endangered ecosystems (Janzen, 1988). They are made more so by periodic droughts and the risk of 'desertification' (FAO, 1997a) from climate induced land degradation through deforestation for agricultural uses and fuelwood collection. The point has to be made that dryland environments, of which 70 percent are already affected by desertification, are inhabited by a large proportion of people who are among the poorest of the world (FAO, 1997a). Thus, people are most responsive to growing a cash generating crop such as tobacco.

It is the merger of social vulnerability, ecological fragility and precarious inherent features of the plant's growing and processing that sum up to the environmental criticality of tobacco growing areas. As a matter of fact, Steinlin (1994) noted that from 1981 to 1990 a trend has emerged in that annual deforestation has progressed more in the uplands (-1.1%) than in the lowlands (-0.8%), while it is lowest in the rain forest zone (-0.6%) where next to none tobacco is grown. And, in the zones most affected by deforestation, either population density is highest, such as in the dry parts of the lowlands, or population growth is highest, such as in the uplands.

The tobacco industry's position on the issue outlined remains somehow opaque. On the one hand, in industry commissioned papers it is recognised that 'the burning of fossil fuels or wood for tobacco inevitably contributes to the greenhouse effect and global warming' (ITGA, 1996c). While, on the other hand, previous acknowledgement of the problem such as raised by Fraser (1986) is sought to repudiate by down-sizing the issue: 'deforestation associated with tobacco curing cannot currently be considered a significant negative externality' (ITGA, 1996a).

From the viewpoint of future research on global environmental change, it is suggested that, due to its socio-economic importance and potential of ecological impact, tobacco deserves well to be integrated in the science plan and activities of several research programmes on global environmental change from a social as well as natural science perspective.

First, integration of tobacco should be done into the International Geosphere-Biosphere Programme's Core Project 'Global Change & Terrestrial Ecosystems' (GCTE, 1998), particularly under focus 3 ('Global Change Impact on Agriculture, Forestry and Soils') and activity 3.1 ('Key Agricultural Systems').

Secondly, making use of the current impetus (or, even, paradigmatic shift) within the global change research community, i.e., from natural to human dimensions of environmental change and towards modelling the earth system by a more region specific and problem oriented approach (e.g., 'transects', 'hot spots', 'fragile environments' or 'regions at risk') (Lohnert and Geist, 1999), tobacco well deserves to become modelled as a major social and economic driver of land-use and land-cover changes in dryland and highland areas of the developing world (e.g., Turner, *et al.*, 1995). It could be seen from a transect study on south-eastern/central Africa that the part of tobacco is so far not well understood. Though next to 90 percent of continental African tobacco is presently grown in countries under dry forest or woodland ecosystems (called 'miombo'), in the respective framework for a terrestrial miombo transect tobacco is only given an appellative mention as a commercial crop, but not explored in full detail (Desanker, *et al.*, 1997).

Notes

1. At the time of paper completion, the merger has only been given newspaper coverage; e.g., *Frankfurter Allgemeine Zeitung* (FAZ), No.9, January 12, 1999, Frankfurt, p.14.
2. The participants of an initiative group of 22 international organisations and individuals, started at the Rockefeller Foundation's Bellagio Study and Conference Center in Italy (June 26-30, 1995), concluded: 'Tobacco is a major threat to sustainable and equitable development. In the developing world tobacco poses a major challenge, not just to health, but also to social and economic development and to environmental sustainability'; Bellagio Group (1995), 'Bellagio Statement on Tobacco and Sustainable Development', *Can Med Assoc J*, vol.153(1), pp.109-10.
3. Curing means the transformation of green leaf tobacco into a pre-manufactured good, mostly done on the farm. Among the four main methods of curing, natural curing makes use of the natural variations in temperature and humidity to dry up the leaves (air-/sun-curing), while artificial curing uses heat from energy sources such as coal, oil, gas and wood (flue/fire-curing). In the case of flue-curing heated air is passed through the harvested leaves by means of pipes or flues, while in the case of fire-curing wood smoke is introduced during the drying process to produce a dark, smoky-flavoured product. In general, tobacco, when picked as a green leaf direct from the plant's stem, must be cured to obtain the characteristic tobacco taste, aroma and colour, and to preserve it for storage, transport and further processing.
4. A compilation and production of data was done since the source of Table 9.1 as specified by ITGA (1996), *Tobacco & The Environment*, Tobacco Briefing, April, East Grinstead, p.2, could not be verified as given (USDA/FAS (1994), *Tobacco Circular FT6*, Washington).
5. It has to be noted that one stack of wood is one metre long by one metre wide by one metre high having a total volume of one cubic metre. Due to irregular gaps and air spaces, only about 60 to 70 percent of the volume is made up of solid wood, so that the weight of wood will range between about 250 to 600 kilograms. This translates into a mean stacking factor of 425 (kg) or 0.43 (tons). While the (mean) conversion ratio of stacked wood into solid wood is 0.43, an equivalent ratio of 2.33 could be used to convert solid wood into stackwood; Fraser, A.I. (1986), *The use of wood by the tobacco industry and the ecological implications*, International Forest Science Consultancy, Edinburgh, p.37.

References

Abedian, I. (1998), 'The Optimal Policy Mix for Tobacco Control: A Proposed Framework', in I. Abedian, R. van der Merwe, N. Wilkins and P. Jha (eds.), *The Economics of Tobacco Control: Towards an Optimal Policy Mix*, Applied Fiscal Research Center, University of Cape Town, pp.5-14.

Akehurst, B.C. (1981), *Tobacco*, Longman, London.

Altman, D.G., Levine, D.W., Howard, G. and Hamilton, H. (1996), 'Tobacco Farmers and Diversification: Opportunities and Barriers', *Tobacco Control*, vol. 5, pp.192-8.

Andreae, B. (1981), *Farming, Development and Space: A World Agricultural Geography*, Gruyter, Berlin, New York.

Blaikie, P. and Brookfield, H. (1994), 'Common Property Resources and Degradation Worldwide', in P. Blaikie and H. Brookfield (eds.), *Land Degradation and Society*, Routledge, London, New York, pp.186-196.

Chapman, S. (1994), 'Editorial: Tobacco and Deforestation in the Developing World', *Tobacco Control*, vol. 3, pp.191-93.

Desanker, P.V., Frost, P.G.H., Justice, C.O. and Scholes, R.J. (1997), *The Miombo Network: Framework for a Terrestrial Transect Study of Land-Use and Land-Cover Change in the Miombo Ecosystems of Central Africa, Conclusions of the Miombo Network Workshop, Zomba, Malawi, December 1995*, IGBP Report No. 41, International Geosphere-Biosphere Programme, International Council of Scientific Unions, Stockholm.

Food and Agriculture Organization of the United Nations (1997a), *State of the World's Forests 1997*, FAO, Rome.

Food and Agriculture Organization of the United Nations (1997b), *Production Yearbook 1996*, vol. 50, FAO Statistics Series No. 135, FAO, Rome.

Fraser, A.I. (1986), *The Use of Wood by the Tobacco Industry and the Ecological Implications*, International Forest Science Consultancy, Edinburgh.

Geist, H. (1997a), *Tobacco Growers of Songea District and their Miombo Environment (Ruvuma Region, Southern Tanzania): Preliminary Results*, Institute of Geography, University of Düsseldorf.

Geist, H. (1997b), *Tobacco Growers of Namwera RDP/Malawi, A Study of Social and Environmental Change: Analytical Draft Report*, Institute of Geography, University of Düsseldorf.

Geist, H. (1998a), 'How Tobacco Farming contributes to Tropical Deforestation', in A. Iraq, R. van der Merwe, N. Wilkins and P. Jha (eds.), *Economics of Tobacco Control: Towards an Optimal Policy Mix*, Applied Fiscal Research Center, University of Cape Town, pp.232-44.

Geist, H. (1998b), 'Tropenwaldzerstörung durch Tabak: Eine These erörtert am Beispiel afrikanischer Miombowälder', *Geographische Rundschau*, vol. 50 (5), pp.283-290.

Geist, H. (1998c), 'Das Bergland von Namwera: Eine Fallstudie über Land-degradierung, Gemeinheitsteilung und braunes Gold', *GAIA: Ecological Perspectives in Science, Humanities and Economics*, vol. 4 (in press).

Geist, H. (1999a), 'Soil Mining and Societal Responses: The Case of Tobacco in the Highlands of Songea (Tanzania) and Namwera (Malawi)', in B. Lohnert and H. Geist (eds.), *Global Environmental Change and Endangered Ecosystems in the Developing World: The Social Perspective*, Ashgate, Aldershot (in press).

Geist, H. (1999b), 'Global Assessment of Deforestation related to Tobacco Farming', *Tobacco Control*, vol. 8(1) (in press).

Global Change & Terrestrial Ecosystems (1998), *Global Change Impact on Agriculture, Forestry and Soils: GCTE Focus 3 Implementation Plan (January 1998 update)*, International Geosphere-Biosphere Programme, GCTE, Stockholm.

Goodland, R.J.A., Watson, C. and Ledec, G. (1984), *Environmental Management in Tropical Agriculture*, Westview Press, Boulder.

Goodman, J. (1995), *Tobacco in History, The Cultures of Dependence*, Routledge, London, New York.

Gossage, S.J. (1997), *Land Use on the Tobacco Estates of Malawi, Report of the Land Use Survey of Tobacco Estates in Malawi 1996*, Estate Land Utilisation Survey, Lilongwe.

Hammond, R. (1998), *Addicted to Profit: Big Tobacco's Expanding Global Reach*, Essential Action, Washington.

International Tobacco Growers' Association (1995), *Tobacco & The Environment*, Tobacco Briefing, April, ITGA, East Grinstead.

International Tobacco Growers' Association (1996a), *Deforestation and the use of wood for curing tobacco*, Tobacco Growers' Issues Papers No. 5, ITGA, East Grinstead.

International Tobacco Growers' Association (1996b), *Tobacco: A major World Crop*, Tobacco Growers' Issues Papers No. 2, ITGA, East Grinstead.

International Tobacco Growers' Association (1996c), *Curing Tobacco: The Issues*, Tobacco Growers' Issues Papers No. 6, ITGA, East Grinstead.

International Tobacco Growers' Association (1997), *The Use of Woodfuel for Curing Tobacco*, Tobacco Growers' Issues Papers No. 11, ITGA, East Grinstead.

Janzen, D.H. (1988), 'Tropical Dry Forests: The most endangered major Tropical Ecosystem', in E.O. Wilson (ed.), *Biodiversity*, National Academy Press, Washington, pp. 130-37.

Kamvazina, S.S. (1994), *Guide to Agricultural Production in Malawi 1994/95-1995/96*, Malawi Government, Ministry of Agriculture and Livestock Development, Lilongwe.

Kweyuh, P.H.M. (1994), 'Tobacco Expansion in Kenya: The socio-ecological Losses', *Tobacco Control*, vol. 3, pp.248-51.

Lohnert, B. and Geist, H. (1999), 'Introduction', in B. Lohnert and H. Geist (eds.), *Global Environmental Change and Endangered Ecosystems in the Developing World: The Social Perspective*, Ashgate, Aldershot (in press).

Mittawa, G.I. (1985), *Tobacco Management Handbook, Notes for Field Extension Staff*, Ministry of Agriculture, Extension Aids Branch, Lilongwe.

Muwanga-Bayego, H. (1994), 'Tobacco Growing in Uganda: The Environment and Women pay the Price', *Tobacco Control*, vol. 3, pp.255-56.

O'Kelly, M. and Bryan, D. (1996), 'Agricultural Location Theory: von Thünen's Contribution to Economic Geography', *Progress in Human Geography*, vol. 20 (4), pp.457-75.

Pater, S. (1994), *Tabak: Rauchsignale auf dem Weltmarkt*, Entwicklungspolitik Materialien No. 34, Deutscher Gewerkschaftsbund, Düsseldorf.

Read, M.D. (1996), *The Politics of Tobacco: Policy Networks and the Cigarette Industry*, Avebury, Aldershot, Vermont.

Reemtsma (1995), *Tobacco: Driving Force for Economic Development*, Reemtsma Cigarettenfabriken GmbH, Corporate Affairs, Hamburg.

Spada, H. and Scheuermann, M. (1998), 'The Human Dimensions of Global Environmental Change: Social Science Research in Germany', in E. Ehlers and T. Krafft (eds.), *German Global Change Research 1998*, National Committee on Global Change Research, Bonn, pp.71-91.

Steinlin, H. (1994), 'The Decline of Tropical Forests', *Quarterly Journal of International Agriculture*, vol.33, pp.128-37.

Tobacco Research Institute of Malawi (1994), *Malawi Burley Tobacco Handbook*, TRIM, Lilongwe.

Tucker, D. (1982), *Tobacco: An International Perspective*, Euromonitor Publications, London.

Turner, B.L. II, Kasperson, J.X., Kasperson, R.E., Dow, K. and Meyer, W.B. (1995), 'Comparisons and Conclusions', in J.X. Kasperson, R.E. Kapserson and B.L. Turner II (eds.), *Regions at risk, Comparisons of threatened environments*, United Nations University Press, Tokyo, New York, Paris, pp.519-86.

Turner, B.L.II, Skole, D., Sanderson, S., Fischer, G., Fresco, L. and Leemans, R. (1995), *Land-Use and Land-Cover Change, Science/Research Plan*, IGBP Report No. 35, HDP Report No. 7, International Geosphere-Biosphere Programme, International Council of Scientific Unions, Human Dimensions of Global Environmental Change Programme, International Social Science Council, Stockholm, Geneva.

United Nations Conference on Trade and Development (1995), *Economic Role of Tobacco Production and Exports in Countries Depending on Tobacco as a Major Source of Income*, UNCTAD/COM Study No. 63/GE.95-51627, Geneva.

United States Department of Agriculture, Foreign Agricultural Service (1994), *Tobacco: World Markets and Trade*, Tobacco Circular Series FT-9407, USDA, FAS, Washington.

United States Department of Agriculture, Foreign Agricultural Service (1997), *Tobacco: World Markets and Trade*, Tobacco Circular Series FT-9707, USDA, FAS, Washington.

Waluye, J. (1994), 'Environmental Impact of Tobacco Growing in Tabora/ Urambo, Tanzania', *Tobacco Control*, vol. 3, pp.252-54.

Williams, M. (1994a), 'Forests and Tree Cover', in W.B. Meyer and B.L. Turner II (eds.), *Changes in Land Use and Land Cover, A Global Perspective*, University Press, Cambridge, pp.97-124.

Williams, M. (1994b), 'The Relations of Environmental History and Historical Geography', *Journal of Historical Geography*, vol.20(1), pp.3-21.

Part 2 – Territorial marginalization

10 The concept of territorial marginality:
A reflection from the standpoint of the geographic image of Portugal at the end of the 20th century

JOÃO LÚIS FERNANDES

Difficulties with the concept of territorial marginality

The question of territorial marginality has recently gained a certain prominence in the domain of Geography. This debate has been broadened to the same extent that the complexity of territorial organisation has been highlighted. Territories today are more interdependent and references are more visible and better known. Inter-territorial hierarchisation nowadays characterises global Geography (Hadjimachalis, 1995). However, despite this favourable climate for discussion about marginality and, particularly in the case of Geography, of territorial marginality, certain basic issues are still to be determined. The concept of marginality is ambiguous and hard to concretise. When we attach the epithet 'territorial' to this term our task becomes even more difficult. There is no unanimity concerning the concept of territorial marginality and, although geo-economic tools predominate, yet there is no concerted opinion with respect to the criteria to be used for a concrete demarcation of these areas. In general, the criteria that unequivocally mark the different degrees of territorial development are complicated. Individual options are important here, since the reading of territories is, very often, a personal task that neither could, nor should be generalised. Territorial marginality is a point of view, a personal way of feeling, before it is a concept in the academic sphere.

Territorial marginality and the construction of the world system

The concept of territorial marginality must have a framework, a set of parameters into which we can introduce the object to be studied. This framework should be based on the vertical integration of geographical scales of analysis (Leimgruber, 1993). We are living at the end of a millennium conspicuous for the intensification of flows of global interrelationship (Dollfus, 1990, 1998). This dynamic, usually identified as globalisation, is a structural phenomenon, which has characterised the identities of populations and places at world level. Many of these flows that sustain globalisation are immaterial, but all have a direct impact on the organisation of territories. Globalisation is also a territorial phenomenon.

The increment of interdependencies has led to the inequitable distribution of relationships of economic and political power, since this framework implies associated dynamics such as selectivity (between the more able, and the less effective), concentration (in the points or areas of greatest potential, a determined function) (Fitoussi and Rosanvallon, 1997). A result of this has been the hierarchisation of territories and, in addition, of professional, social and other groups (Delapierre, 1995).

The shaping of the World System, a foundation of the globalisation process, is a privileged scheme for the analysis of inter-territorial hierarchies at the end of the 20th century. The marginality or centrality of a territory depends largely on the functionality of each unit, the way in which each territory fits into a competitive world and on the functions, rarer or banal, of greater or lesser value, that constitute the arguments of a territory in a world system organised in multiple networks (Figure 10.1). A classification of an geo-economic nature predominates in this contextualization, in a scheme that should be interpreted flexibly. To recapitulate, we have a perception of territorial because, in the first place, centres exist and, in the second, because we have a system of reference. This is also why marginality is today a theme that is more discussed, more visible, more perceived.

Position of Portugal in the world system

Portugal is an example of how difficult it is to consider the positioning of territories in the international context. This difficulty is the result not only of the complexity that characterises the country in contemporaneity, but also of the geo-historical course it has followed, especially from the time when we found ourselves at the beginning (always undefined) of the global system that has become the framework of world societies at the end of the 20th century. This complexity is not exclusive to Portugal. There are many cases that are very hard to classify, which simply serves to reinforce the

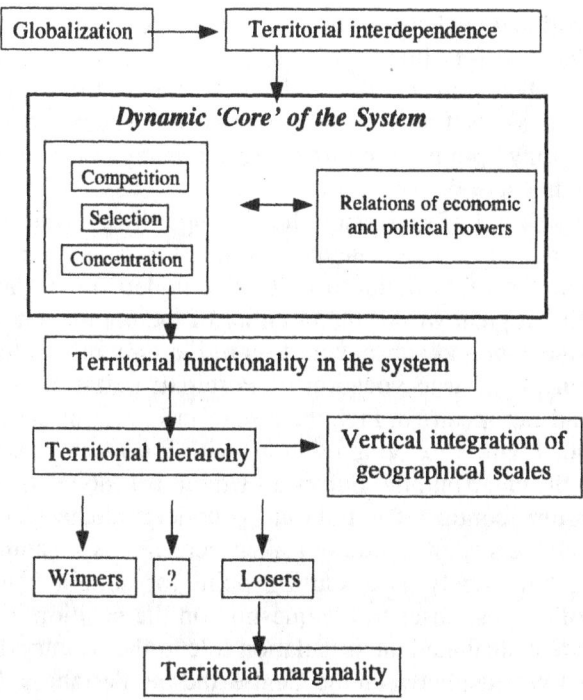

Figure 10.1 Globalization and territorial marginality

view that the scheme described earlier has to be analysed as flexibly as is expediently possible, in a World of cultural, historical and geographical heterogeneity.

All this makes reflection on the relative position of Portugal in the world context difficult, for two main reasons: we are not dealing with a homogeneous unit, nor will the national scale, very probably, be the most suitable for this exercise. However, despite the crisis of the Nation-State, national analysis must be a constant.

It is common to say that the most glorious moment in the History of Portugal, with repercussions on its Human Geography, were the Maritime Discoveries. Indeed, after a period of Social and Economic History of a Humanity built on a base of different World Economies, with the slightest of connections among them, the Maritime Expansion of the Iberian peoples and maritime powers such as England and Holland corresponded to a new phase of broadening cultural and economic horizons and reinforcing interdependency among the civilisations that comprise the mosaic of humanity. Portugal has helped to construct the World System; it has taken part in building a world that is closer and smaller. This fact, regardless of the consequences it has had for Humanity, noted with greater or lesser nationalistic enthusiasm, distinguished the 16th and 17th century realities of the country, as well as its future. Contemporary Portuguese identity and its Human Geography (population structure, for instance) owe much to its profile during this period.

It is to this period of expansion that Portugal owes some of the myths that still feed the Lusitanian imaginary today: the construction of an empire; the expansion of Portuguese culture, founded on its language; the contact with the tropical world; the mobility of the population, in a tireless colonising movement and subsequent loss of the empire. All this brings us to the question of the true vocation of Portugal: seafaring and Atlantic? Continental and European? In fact, the loss of influence in the overseas territories coincided with improving relations with European, first with EFTA and then with the entry into the European Union in 1986 (Ferreira, 1998).

In an enquiry conducted on about 80 undergraduate Geography students by the University of Coimbra (taken here not as a sample of a particular reality, but merely as a starting point for certain considerations) around 90% of the responses to the question on the position of Portugal in the World System attributed an articulation role to the country: between the First and Third Worlds; between the Centre and the Periphery; between the economic and political powers and the fringes of decision-making; between Africa (and the Portuguese-speaking world in general) and Europe; between the Atlantic dimension of our identity and its continental aspect.

This response also has its roots in the hostile and, in many ways, contradictory social and economic development of the country. The contradiction is reinforced when we look at Portugal's heterogeneous capacity for response in the context of international integration. 'Intermediate development', 'national design' of articulation, to quote the reflection of a Portuguese politician (Mário Soares), 'historical role of articulation' and 'intermediate geographical position' in a Euro-centrist *planisphere*; these expressions were used most often in the justification of the response to this question. This characteristic of articulation was further seen as a potentiality. In an open system, the main assets are in the geo-economic Centre, but they are also in areas which, though not central, are taken to be areas of contact, in the context of which *lusofonia* [Portuguese-language-using], (the Portuguese cultural frontier, as Adriano Moreira has said).

From this batch of responses we would further highlight the argument 'design of the past', which coincides with a Historical Geography of Portugal, well embedded in the Portuguese consciousness. The actual position of the places in the system is rooted in the system itself, as well as in the identity of each territory and in its course throughout history.

A country's past is very important to its present placement in the world geo-system. Thus the international position of Portugal is conditioned by the past, by the successes and failures of preceding generations, but any kind of historical determinism should be profiled as clearly as possible. It is better to set this reflection in another domain – in that of temporal cycles. Indeed, the international position of Portugal is a structural matter, but this does not prevent the attribution of some responsibility to it at more conspicuous and significant conjunctural moments.

Approximately 15% of the respondents mentioned the period of Portugal's formation in the geopolitical context of the Iberian Peninsula as one of the most important conjunctures for the future of Portugal in the global context. Over 40% of the responses focused on the 'Discoveries–April Revolution–De-colonisation trilogy'. Nearly 30% of respondents stressed the importance of membership of the EU. In fact these figures are a fairly faithful reflection of some of the myths that feed the Portuguese imaginary, and some of the periods that are most firmly fixed in the national consciousness. These results have gained further interest inasmuch as they underscore Portugal's placement in the international system positively, within the framework of the much-mentioned geometric variable (Iberian Peninsula-Africa and Portuguese-speaking World-Europe). These options are interwoven in a territorial logic of Expansion-Retraction and Opening-Closure: first, in the Iberian Peninsula, with the establishment of Portuguese sovereignty; after expansion, construction of an Empire and its disappearance; finally, Portugal's opening up to the European space. The placement of territories within the system and the concepts of centre and periphery scattered around spheres of different domains, but constituting, above all, a territorial issue of interdependence, expansion and shrinking.

The placement of territories in the international context is related to their interconnection with other territories, with the way this relationship is designed and, equally important, with the introduction of the time factor in this logic: how are these interrelationships proceeded with during the course of historical transformation? Interrelations that are it in terms of economic power, political power and sovereignty/dependency.

The frontiers of development in mainland and island Portugal

Portugal today is an open country. This openness took a new form with European integration and the productive restructuring of Europe in the context of consolidating the Economic and Monetary Union. Portugal has also always been a country with development divisions. A national study, the relevance of which is undisputed, should be associated with a reflection on a more detailed scale.

This was the challenge passed to the students surveyed. They were requested to look more closely at Portuguese Geography. Before attempting to define the marginal and central areas, the respondents had to outline on map (Figure 10.2) the territories they classified as peripheral.

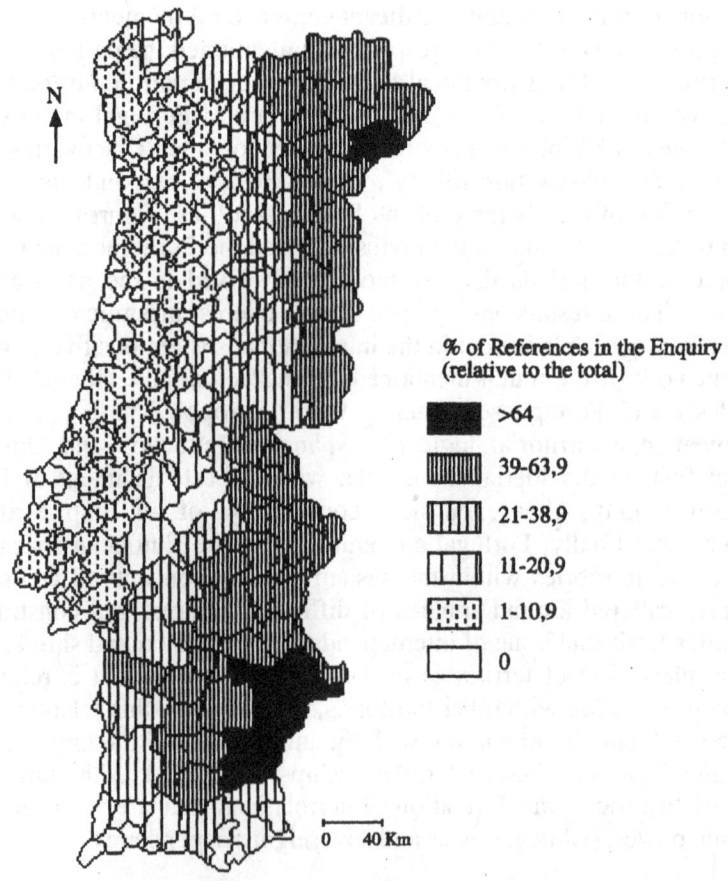

Figure 10.2 Marginal territories in Portugal, according to the enquiry

This result is translated into a schematic view of development of the Portuguese territory and assimilation of concepts such as 'coastality' and 'interiority'. Besides, this result testifies to the consequences of history, particularly of Portuguese maritime expansion. So the map constructed from the responses to this question virtually establish a limit between the Interior, largely regarded as marginal, and the Coast, characterised and identified as central. The picture presented, therefore, translates the perception that these students have of Portugal and the way that most Portuguese look at the Portuguese territory. The coast is characterised by its proximity to the Atlantic, as a synonym of prosperity; the interior, closer to the Spanish border, is viewed as deprived area.

Winning areas, as opposed to losers. Asked about synonyms for the concept of marginality (Table 10.1), those questioned showed a pessimistic outlook on the real nature of the areas referred to as marginal. According to this enquiry, the interior thus signifies extreme pessimism.

Table 10.1   Terms associated with the concept of marginality

Exclusion	Discrimination	'At the side'
Peripheralism	Solitude	Oppressed
Under-development	Disintegration	Individualisation
Isolation	Different	Inferiority
Criminality	Segregation	Decadence
'Set apart from....'	Minority	Abandoned
Poverty	Backwardness	Disinterest
Repulsion	Inaccessible	'Be overlooked by...'
'At the edge of...'	'Not included...'	External
Remoteness	'Different from...'	'Lack of...'
Racism	Unattractive'	Agriculture
Inequality	Unprepossessing	Shanty towns/ shacks
Limit	Dependency	Discontinuity
'Outside...'	Corruption	Depression
Non-adaptation	Indifference	Misery
Separation	Corner	Disjointed
Neglect	Asymmetry	Interiority
Rejection	Disparity	Distance
Disregard	Heterogeneity	Quality of life
	'On a second level'	Precariousness
	Exterior	Unsociability

This view, therefore, matches the images of our country in cartographic representation and geo-demographic (or geo-economic) indicators, such as the Ageing Index, Population Density, and Migratory Balance, among others. Even so, the image of the country changes when we produce evidence from other territorial readings. The same respondents were asked to indicate on a map the territories (in this case, municipalities) that they felt enjoyed the best quality of life (Figure 10.3), and the municipalities where they would like to live in the future (although this map would be conditioned by the actual place of residence and place of birth of the respondents) (Figure 10.4). In this case, the responses are not so schematic, denoting a wider territorial view, and one that is more diffuse.

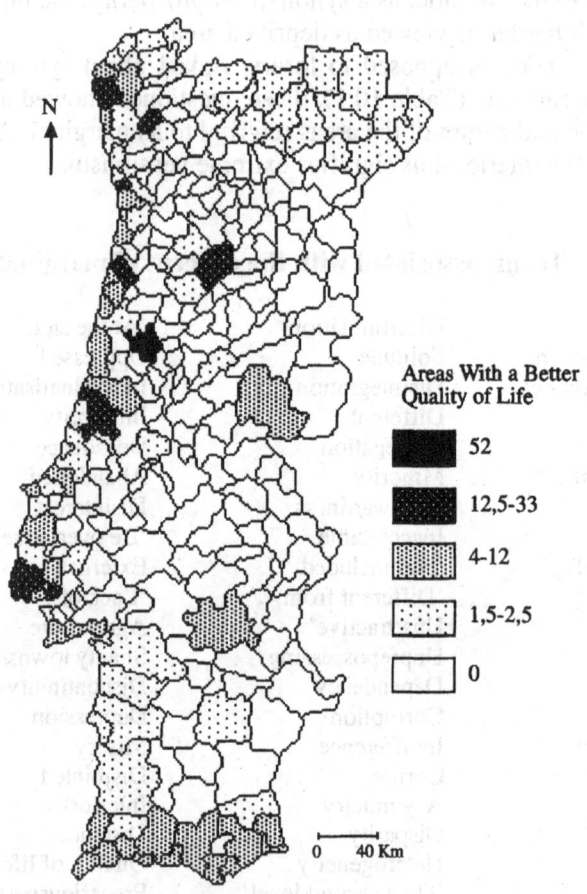

Figure 10.3   Municipalities with a better quality of life, according to the enquiry

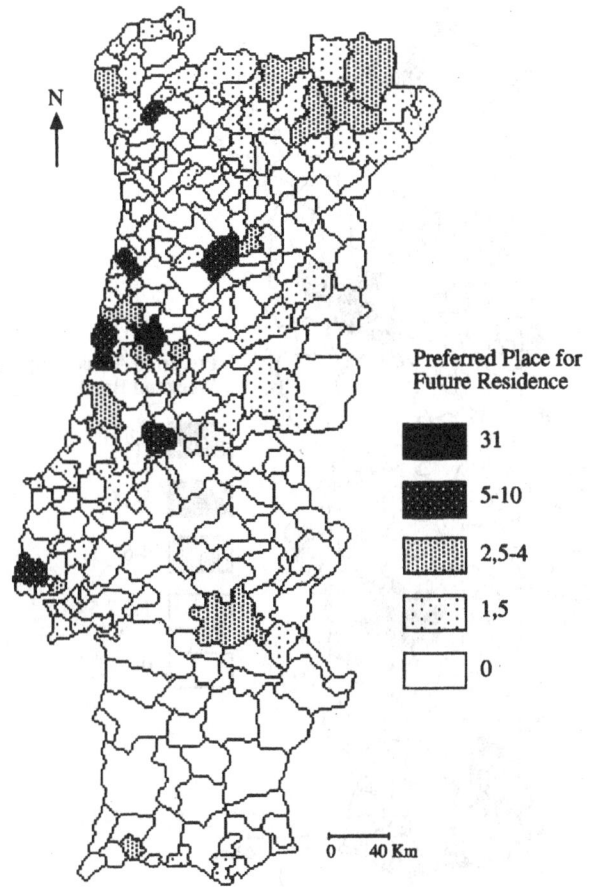

Figure 10.4 Municipalities preferred for future residence, according to the enquiry

The maps presented in Figures 10.3 and 10.4 do not exhibit such rigidity as in Figure 10.2. The relation between marginality and concepts such as quality of life and preferred daily lives (because it is this logic that is tackled by the question on preferred place of residence) is extremely complex.

Yet, despite the negative connotation of these territories, the response to the question which asked for the choice of a municipality (or several) to spend a weekend break (Figure 10.5) was interesting. The rigid coast/ interior perception/division is completely cast into the shade and, on the

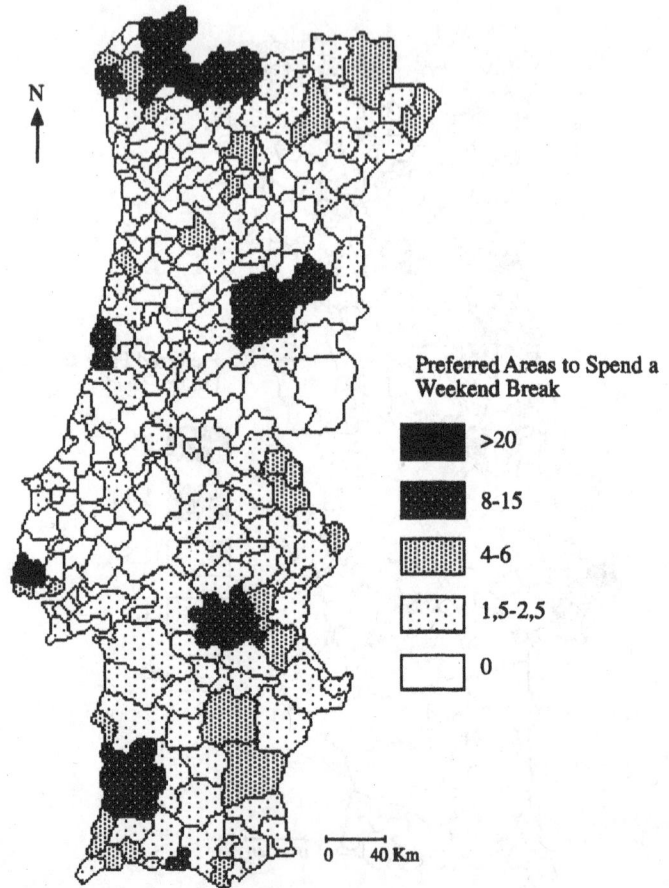

Figure 10.5 The best municipalities to spend a weekend break, according to the enquiry

contrary, some of the most mediated areas in the national territory move into the limelight: mountainous areas, like the Parque Nacional Peneda-Gerês and the Serra da Estrela; Trás-os-Montes, and the Alentejo territories consistently regarded as marginal in Figure 10.2).

These choices were explained by factors more usually associated with the stratification of territories with different levels of development, such as the existence (or not) of thriving businesses, the relative position with respect to important communication routes and/or to structuring urban centres.

The existence and quality of services and, in general, the positioning of places in relation to the principal flows (material and immaterial) that accompany the opening of spaces are other considerations. However, the responses still insisted on factors of a more qualitative nature, sensory and sometimes difficult to interpret. Among these were conservation of heritage and environmental equilibrium; harmonious inter-personal relationships; the safeguarding of a healthy daily life. In this sphere, territorial symbols are more highly valued; the beach, the mountains, the rural world in general. Specific localities were mentioned, like the *Barragem da Caniçada* (Caniçada Dam) in Terras de Bouro (District of Braga): 'There can't be anything more wonderful than to live two steps from the dam at Caniçada', according to one respondent.

Given all this, are we in the presence of new paradigms of looking at and feeling about the territories? New behavioural codes that will be translated into new geographies? Or will these maps simply reveal the intention to attribute functionality to marginal geographies; in this case, tourism? It will take more refined research to explore this line of thought.

In conclusion

As a summarising conclusion it is possible to say the following:

1. The concept and real meaning of marginality is a complex reality.
   The analysis of the dynamics flowing through geographies of marginality is a far easier task and globalisation is an appropriate framework to accomplish it, and interdependence/functionality/hierarchy constitute the Dynamic Core of the System.
2. What kind of functionality does each territory have in the world system?
   Marginality and centrality depend on functionality and they depend on the way the territory will achieve good or bad integration in the world geosystem.
3. The image obtained is not static, but a dynamic, since this reality is the result of history.
4. In all reflections on marginality we have to integrate different geographical scales vertically.
5. The territorial perceptions of ordinary people are strategic questions in this kind of thinking, since 'marginality' is not just an academic issue.
6. This perception testifies, above all, to an geo-economic interpretation of the territory, and a pessimistic view of marginality.
7. In spite of this, there are profits to be found in marginal areas.

We are dealing with different paradigms of development, and a new behaviour code. In this context, individual readings of this problem are important. In the end, it is the populations, composed of individuals, which are the principal agents that direct, structure, operate and develop their everyday lives in the territories in which they live; in the ambit of an attempted definition of marginality, no academic consideration on the territory can ignore this fact.

References

Delapierre, Michel (1995), 'De l'internationalisation à la globalisation' in Michel Savy and Pierre Veltz (eds.), *Economie Globale et Réinvention du Local*, DATAR/Éditions de l'Aube.
Dollfus, Olivier (1990), 'Le système monde' in *l'Information Géographique*, 1990:54, Armand Colin, Paris.
Dollfus, Olivier (1998), *A Mundializaçao*, Publicaçoes Europa-América, Lisboa.
Ferreira, José Medeiros (1988), *Posiçao de Portugal no Mundo*, Portugal, Os Proximos 20 Anos, 1988: IV, Fundaçao Calouste Gulbenkian, Lisboa.
Fitoussi, Jean-Paul, and Pierre Rosanvallon (1997), *A Nova Era das Desigualdades*, Editora Celta, Oeiras.
Hadjimichalis, Costis and Sadler David (1995), 'Integration, Marginality and the New Europe', in Costis Hadjimichalis, and David Sadler (eds.), *Europe at the Margins. New Mosaics of Inequality*, John Wiley and Sons, Chichester.
Leimgruber, Walter (1993), 'Marginality and Marginal Regions: Problems of definition' in David Chang *et al.* (eds.), *Proceedings of the Study Group on Development Issues in Marginal Regions*, National Taiwan University, Taipei.

11 Territorial positionament of a peripheral region in the Atlantic Arch, Galicia (Spain)

MARÍA JOSÉ PIÑEIRA MANTIÑÁN AND ROMÁN RODRIGUEZ GONZÁLEZ

Introduction

Galicia, with a population of approximately 2.7 million inhabitants and an area close to 30,000 square kilometres, is one of the Autonomous Communities of Spain. Its location on the occidental end of Europe (Figure 11.1) confers on it its two major characteristics as a regional space; those of being Atlantic and peripheral. Its Atlantic nature may be perceived in the marks left by the Ocean, non only on the environment but also on the successive patterns of human occupation throughout history. At the same time, Galicia's peripheral character arises from its situation in the north-western corner of the Iberian Peninsula – one of the most occidental areas in Europe itself – along with its traditional difficulties in terms of accessibility (Puyol Antolin and Vinuesa Angulo, 1995). This marginal location also affects Galicia's low level of relative development in the context of the European Union (Figure 11.1).

In this sense, Galicia is a disadvantaged region that belongs to the administrative level NUTS 2 and is included in the Objective 1 for issues of Regional Policy (see Comisión Europea, 1995a). Its GDP, considered in relation to the purchasing power, amounts to 59% of the European average and 83% of the national average (see Comisión Europea, 1995b; IDEGA, 1998). Despite this comparative position, the most recent evidence suggests a strong modernisation and urbanisation process in Galician society and its productive and territorial structures. The level of agrarian employment nowadays is not relevant in relation to the tertiary sector of the labour market. There has also been an improvement in industry and a generalisation of new collective behaviours (mobility, monetarist processes, and role segmentation), derived from the consolidation of the market capitalist

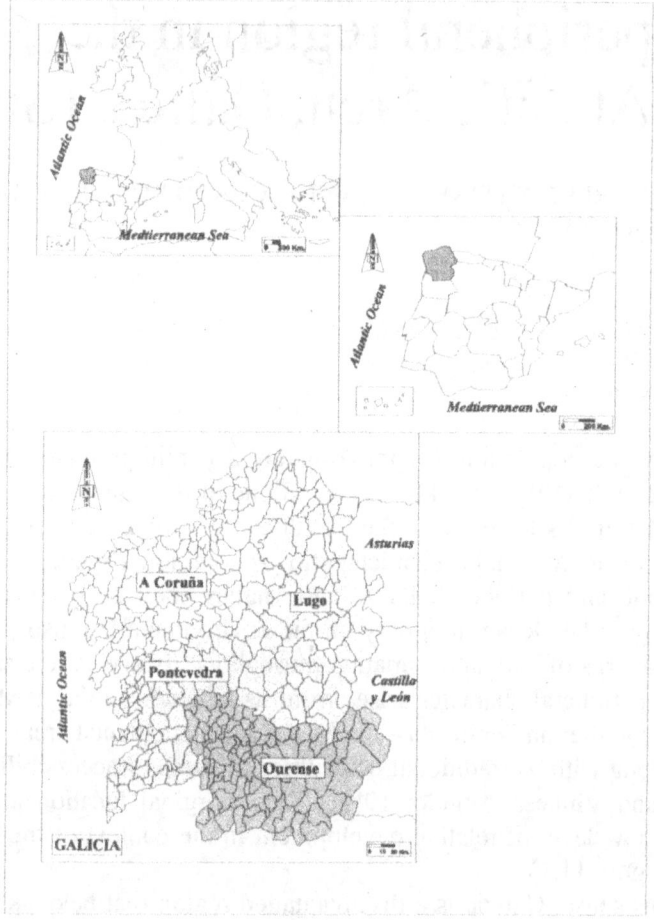

Figure 11.1 Galicia's location in Europe and in the Iberian Peninsula, provincial and municipal division of Galicia

economy within the territory. In this sense, Galicia's present economic situation differs quite broadly from its former condition in the 1950's as a basically rural agrarian region; today agrarian employment, in spite of being one of the highest in Europe, does not reach the 20% level. (It was about 70% in the 1960's.) In addition to this, almost 50% of the population live in settlements of more than 50,000 inhabitants, a double percentage in respect to that of previous decades (e.g., A. A. V. V., 1994; Comisión Europea 1995b).

A clear indicator of the peripheral situation is the fact that Galicia has been a traditional source of emigrants. Since the 1960's more than a million people are estimated to have taken part in migratory processes towards European countries such as France, Germany, and towards more industrialised Spanish regions, such as the Basque country, Catalonia or Madrid. The effects of emigration are evident in the present demographic structures, with a significant ageing of the population. Emigration has also affected Galicia's birth rate (the lowest rate in the European Union today, 8‰) and has caused the depopulation of rural areas, the main sources of emigration.

Galicia's Atlantic idiosyncrasy is logical if we bear in mind the extension of its coastal line - 1,195 km, a third of the Spanish coast - along with the penetration of climatic influences inland. The Galician coast is quite rugged and irregular, with excellent natural ports. From a historical point of view, this situation has favoured the existence of a significant human occupation linked to the exploitation of the marine resources and an intensive agriculture with a high yield. (During the past years new resources, like tourism, have originated). For this reason, an important number of settings have been consolidated in this territorial context; along with a dense network of small cities with diversified activities, the main Galician cities are located in coastal and port areas.

Until recently, the Atlantic Ocean was also the main channel of communication of this region with the rest of the world. Innovations and novelties from other countries were introduced through its ports. At the same time, numerous Galician individuals set out from them for their transoceanic emigration. The ports were also the centres of commercial business and industry. They were, in short, the 'doors' of Galicia. These circumstances were encouraged by the difficulty of territorial communications (a mountain range separates Galicia from the rest of the Spanish State) and the late development of air transportation in the area.

Bearing these two factors in mind, in the pages that follow we will show how since 1986, after Spain's entrance into the European Union, Galicia has initiated an important territorial and economic geo-strategy. This strategy pursues the reinforcement of its Atlantic character and the reduction of its peripheral nature, with intervention lines that aim at an improvement of Galicia's position in the regional framework of the Union (see CEDRE, 1992; Hildenbrand, 1996). It is for this reason that Galicia's present territorial stance and its development strategies are laid out in terms of its Atlantic character (increasing contacts with Northern Portugal, participation in the Atlantic Arch) and on the basis of the budget provided by the Regional Development Policy of the European Union.

The Atlantic nature of Galicia as a territorial factor

Galicia's model of spatial organisation presents a clear Atlantic character. As a region belonging to the Atlantic Arch it shares problematic, environmental and socio-economic conditions with other European regions (see CEDRE, 1992). Among all of them, Northern Portugal is undoubtedly the region that has more interests in common with Galicia. Economic flows and exchanges between both have increased considerably since 1986. This has been complemented with a cultural and political rapprochement, which tends to strengthen the Atlantic nature of this so-called axial 'Euroregion'. Thus, customs procedures have been liberalised, negotiation possibilities have been initiated, and there exists an increasing transfer of people and goods between both regions (Lois González, n.d.; Lois González and Piñeira Mantiñán, 1998). An example of this new situation may be the recent opening of important Portuguese bank branches in the south of Galicia (Figure 11.2).

The Atlantic character is fundamental in terms of the territorial structure of Galicia. We may perceive (Figure 11.3) a basic contrast between the littoral space and the interior. Whereas the former concentrates two-thirds of the population and 70% of the regional wealth, along a north-south axis, the latter is threatened by an intense loss of population and a primarily agrarian activity. The Atlantic coast, with the exception of certain northern sectors has a high density of urban settlements of different sizes as well as a productive diversity. By contrast, the interior is organised functionally around the provincial capitals and some small town, focal points in the rural space.

The urban network lacks a major leading focus that organises the entire region functionally. Quite differently, we may find a bipolar system centred in two cities with a similar size, A Coruña and Vigo. With a population around 250,000 inhabitants they functions as the main urban points of reference for the regional north and south respectively. Despite the differences regarding their local bases of development, their spatial projection is quite similar. Whereas A Coruña has been traditionally linked to administration and commercial services, Vigo has a more industrial character. However, both cities found part of their vitality and urban features on their port activity; both are in fact leading fishery ports, supporting an important transforming industry (Vigo is the major Spanish port in terms of fish landings). These ports operate as entrance and departure points for industrial goods and raw material (oil in A Coruña, granite and vehicles in Vigo),

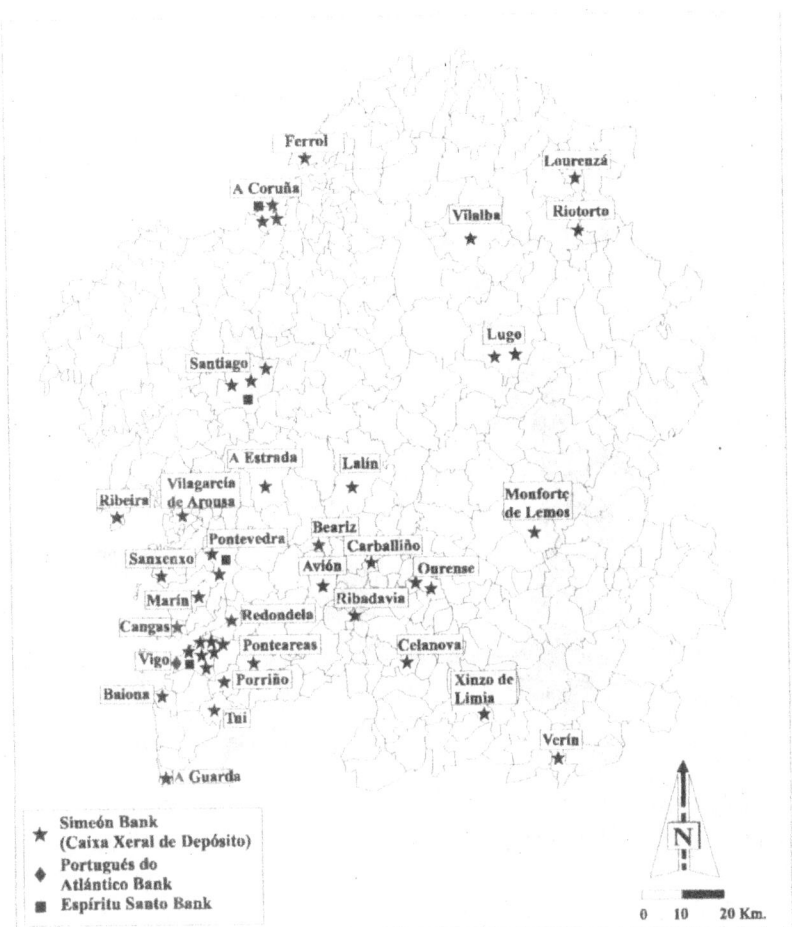

Figure 11.2 Portuguese bank branches in Galicia

and finally, both cities organise around them two modest metropolitan areas – also called urban regions – that contain small urban centres with specialised functions (tourist resorts, industrial areas).

At the same time, A Coruña and Vigo function as extremes of the coastal urban axis, which extends up to the border with Portugal in the south and reaches Ferrol in the north. Ferrol and Pontevedra are small-size cities (around 70,000 inhabitants) integrated in the above-mentioned urban regions. Ferrol's development incentives rest upon the naval sector in

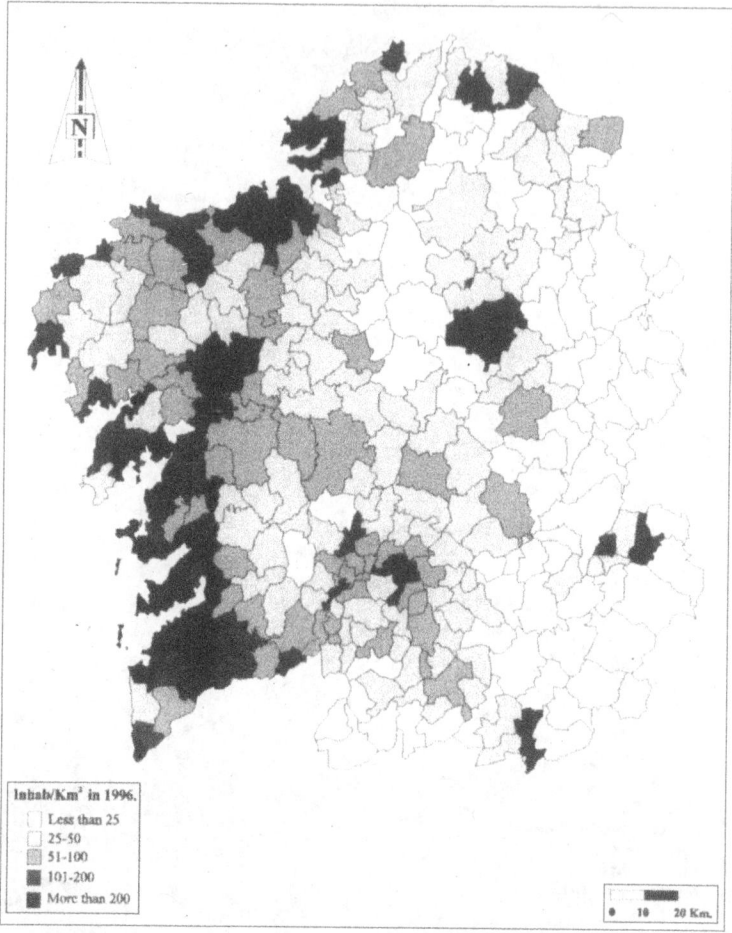

Figure 11.3 Map of population density in 1996

crisis, upon the presence of an important military contingent, and thirdly upon being upon the presence of an important military contingent, and thirdly upon being the driving force of the Ferrolterra region. Pontevedra, in contrast, concentrates diverse administrative services as a province capital.

In the interior Lugo and Ourense stand out as the leading cities regarding industry and services within their respective provincial territories. On the other hand, Santiago de Compostela appears as the balance point between Vigo and A Coruña within the urban structure of Galicia. This city fulfils a series of functions at a regional and international level (university town, regional capital, tourist centre, and religious pilgrimage destination), that project its image and influence highly above its demographic volume (nearly 100,000 inhabitants).

In other words, Galicia's territorial occupation is clearly Atlantic in nature, the coastal urban axis concentrating a major part of the population, resources and facilities (Malosse, 1996). Furthermore, in the present we may perceive an evident interest in extending and connecting this development corridor with the Atlantic facade of northern Portugal (connection through motorway, construction of new bridges).

Galicia's stance and development strategies

Galicia, as a disadvantaged region, organises its main intervention lines and its socio-economic dynamics trough the Regional Development Programmes (see A.A.V.V., 1994; IDEGA, 1998). The program in force comprises the period 1994-1999. It entails a government commitment of action by means of investment plans in the different administrations. The impact of this program and its execution are key elements to consider when weighing up the region's stance and development geo-strategies in the above-mentioned period, since the program affects important areas, such as infrastructure, agrarian and education policies.

The Regional Development Programme in force aims at promoting a sustainable, well-balanced growth of the regional economy. Other primary objectives are the improvement of the population's life quality and the achievement of a greater territorial balance. The Programme seeks, in other words, a greater convergence with the other European regions. This is only a voluntaristic declaration of intentions that must be attained by promoting the region's favourable aspects in terms of development and, obviously, by solving the existing major problems.

The planning for projects with the financial support of the Structural Funds is organised in eight intervention lines. These lines are complemented with others articulated in the Union Initiatives and the Cohesion Funds. Regarding the first group, the chart below shows that the largest expenditure corresponds to *Territorial Integration and Articulation* (TIA), with 64% of the total budget in the Operative Programme (Table 11.1).

Table 11.1 **Intervention axis of the Structural Funds for the period 1994-1999, in millions of pesetas; 166.386 pesetas = 1 €**

	Total cost 1994-1996	E.U. Participation	Other Administration	Private Sector	Level of completion for 1994-96
TIA	254,454	65.5	34.5		91.7%
DEN	63,377	26.1	9.9	64.0	57.1%
T	3,954	32.2	19.2	48.2	48.2%
ARD	3,247	70.3	26.7	3.0	110.2%
F	1,814	41.4	20.5	38.1	133.7%
ISEA	55,620	60.6	39.4		101.4%
VHR	12,070	74.6	25.4		68.4%

This line fundamentally involves the construction and improvement of communication infrastructures. It tends, on the one hand, to facilitate the internal and external accessibility of the region; on the other, it seeks to establish the necessary bases to ameliorate the standard of living and the competitive possibilities, enabling better conditions for enterprise. The objective of these strategies is to counteract the region's peripheral character. The European Union confers this action line a priority degree. It provides, however, 65% of the total cost through the FEDER. The level of execution for the period 1994-1996 is considerably high. With regard to the road network, the strategy concentrates on the improvement of internal communications and on some sections of the highway connecting with the Peninsular plateau. The main consequence would be a greater integration, both at a regional level and within the Trans-European transportation networks. The plan also affects the modernisation of railway infrastructures, construction works in ports and airports, the improvement of media services, and several environmental protection actions.

Another priority intervention line is the one called *Development of the Economic Network* (DEN). During the first two years of the Programme the level of execution of the managing organism, the Galician Institute for the Economic Promotion, has been rather low. This was due to a delay of the European Committee in approving the regulation policies for its management. In this case, there is a strong participation of the private sector. In fact, the beneficiary companies have provided 64% of the estimated costs. The strategy is bases upon the support of small and medium enterprises as a point of departure for entrepreneurial development. The final goal is an improvement in the quality and productivity of companies. This is intended to be achieved by means of business training initiatives, by financial aid for investment interest rates, by the promotion of the technological innovations and others.

Tourism (T), the third intervention line, is strategic for Galicia as a peripheral space with considerable natural, artistic and cultural resources. Similar to the previous case, this line is conceived to be financed jointly by tourist companies and the Administration. Due to the characteristics of the Galician territory, this intervention line centres upon the promotion a of high-quality tourist offer with variety. This involves the restoration of monuments, the rehabilitation of natural spaces, the expansion of outdoor tourism, improvements in the promotion and marketing of the sector, and specially the development of a tourist image associated with the Road of Santiago. This ultimate goal is to achieve a position within the tourist market, attractive enough to generate an increasing number of visitors.

The intervention line of *Agriculture and Rural Development* (ARD) has undergone a level of completion higher than estimated for the first phase of the Programme. The major part of its budget is intended to adapt Galician agriculture to the Community Agriculture Policy with the financial support of FEOGA-Orientation. This is a high-interest line for the European Union, and, as a consequence, it provides more than 70% of the assigned capital. The principal actions are; a reorganisation of the agrarian property (land concentration), an improvement of (agro-food) technology and the quality of products, training for the management of agrarian companies, integration in the production, manufacturing and marketing sectors, and a diversification in the rural activities. As a whole, the plan fosters the maintenance of a reasonable agrarian activity within the region, with productive ends and a desire to preserve the rural population. (Nevertheless, Galician agriculture receives proportionally some of the lowest financial support in Europe with strong restrictions in strategic sectors, such as milk quotas).

Fishing (F), manage through the IFOP, has generated higher expenses than estimated for the period 1994-1996. This was essentially due to the initial outlay, necessary for the renovation and modernisation of the fishing fleet and for the manufacturing and marketing of products. The repercussion of Galicia's entry into the European Community with regard to fishing goes beyond the economic contribution. It affects, as well, the fleet's capacity of negotiation with other countries. Galicia's stance in this respect focuses on the reorganisation of the sector, the development of fish farming, and the improvement of the commercialisation process.

The intervention strategy of Infrastructures for the *Support of the Economic Activity* (ISEA) bears some similarities to Territorial Integration and

Articulation. Both involve a financial contribution, which is exclusively public and comes primarily from the Union. This makes Galicia's lack of infrastructures evident, as a peripheral, low-developed region. The European Union deems it necessary to rectify this situation in order to make a greater economic development of the regional companies possible.

Valuation of the Human Resources (VHR) is considered to be of maximum interest for the Union that provides about 77.8% of the budgeted capital. Its execution centres upon two different actions. The first action consists on the improvement of the quality of schooling through the construction of education centres, teacher training programmes, etc. The second one involves specialising education towards the labour market; occupational courses, professional retraining.

The Cohesion Funds, on the other hand, are intended fundamentally to the construction of communication systems to connect Galicia with the Trans-European system (in particular, the highway Rías Baixas) and to the execution of environmental rehabilitation projects (drainage, hydraulic works). Other investments of great interest are channelled through the European Union Initiative Interreg (those of bridges and the connection of the transport infrastructure with the Portuguese border). These investments are evidence of a clear interest in increasing the relations with northern Portugal.

As a conclusion, the foregoing discussion has suggested that Galicia's stance and development strategies are organised in terms of its integration in the European Union. The financial support of the Union is fundamental for the construction of the infrastructures and the equipment necessary to counteract, as far as possible, Galicia's peripheral nature. Furthermore, the main intervention lines, organised through the Regional Development Programmes, establish the basic initiatives to be followed by the Administrations and by the economic agents that will benefit from the financial support, either directly, with specific projects, or indirectly, through the transformation and improvement of the region in general.

References

A.A.V.V. (1994), *Integración y revitalización regional*, Asociación Castellano-Leonesa de Ciencia Regional, Salamanca.

CEDRE (1992), *Etude prospective des Regions Atlantiques*, Bruselas.?

Comisión Europea (1995), *Europa 2.000+*, Cooperación para la ordenación del territorio europeo.

Comisión Europea (1995), *Galicia en la Unión Europea*, Madrid.

Hildenbrand Scheid, A. (1996), *Política de Ordenación del Territorio en Europa*, Universidad de Sevilla, Sevilla.

IDEGA (1998), *A Economía Galega, Informe 1996/97*, Santiago de Compostela.

Lois González, R, (n.d.), *Galicia-Regiao Norte de Portugal y la posible formación de un espacio económico común en la periferia atlántica. Apuntes para un debate*, Il simposio international articuolación de territorios de la frontiera, Fundación Afonso Henriques, Zamora.

Lois González, R. and Piñeira Mantiñán, M.J. (1998), *Os espacios construidos en Galicia: pervivencias da cultura labrega despois do proceso de urbanización*, Actas del seminario internacional - Cultura y Arqihtectura, Universidad Fernando Pessoa, Porto, Editorial Lea, Porto.

Malosse, H. (1996), *Europa a su alcance*, Fundación Galicia-Europa, Santiago de Compostela.

Puyol Antolín, R. and Vinuesa Angulo, J. (1995), *La Unión Europea*, Editorial Síntesis, Madrid.

12 Reality and perception of marginality:
The case of rural Catalonia

MÁRTI CORS

Introduction

On the threshold of the 21st century, it has become an increasingly complex process to define and classify certain geographical spaces as marginal. This complexity is due, in part, to the fact that in the West marginality has acquired an increasingly mobile nature, gradually severing links with the strictly physical framework of the territory, due to improvements in communications. Today, the concept of marginality is more closely concerned with social and cultural characteristics, which lead to the appearance of new marginal spaces. As occurs in most cases, the perception of a territory is a function of our actual knowledge of this territory. If we accept the general idea that marginal areas are not static (Majoral and Sánchez-Aguilera, 1998), the difference between the reality and the perception of a territory will become greater as its socio-economic dynamism grows.

The degree of marginality of a space is relative and depends on the scale of analysis, which is applied. It is for this reason that the territorial framework, which is to be used, should be the main reference point when evaluating the marginality in the study area. Here, the territorial framework is Catalonia, a region in the extreme Northeast of Spain, on the shores of the Mediterranean Sea. The territorial analysis is centred on the *comarca* of Solsonès, a rural area in the interior of the region.

The study seeks to fulfil a series of specific objectives related with the concept of territorial marginality. The initial idea of this study is to show how certain rural areas in the interior of Catalonia, which had traditionally been considered marginal areas, both in real terms and in terms of the perception of this reality, no longer have this marginal nature. This is, at least, apparent in many of their facets, due, on the one hand, to economic development and the improvement in the infrastructure of these zones and, on the other, to the narrowing of the gap between rural and urban spaces and vice versa, a situation which is related to a new scale of values and social behaviour. In spite of all this, certain stereotypical aspects refuse to be cast off concerning the perceptive value of rural life which means we still have to talk of a certain degree of marginality of a perceptive nature in these areas lying in the interior of the region of Catalonia. The line adopted in this article and the specific objectives of the study are based on a number of articles published by the research group based at the University of Barcelona concerned with marginal areas. This paper has been undertaken very much in line with the various contributions made by the members of this group on the reality and perception of marginality in rural Catalonia (Majoral and Sánchez-Aguilera, 1998; López-Palomeque, 1996; Font and Capella, 1998). As a response to this general approach, we have set ourselves the following objectives, around which our work is structured:

- To analyse the traditional isolation to which the *comarca* of Solsonès has been subjected due to a range of circumstances, including, its geographical location in relation to its relief features. This area's present-day socio-economic characteristics are partly the result of this physical isolation caused by the orography of the territory (mountainous zone). In general, then, the marginality of Solsonès has been territorial in that it is associated with that of a peripheral *comarca* in relation to the main market centres.
- To evaluate the present-day reality of the *comarca* using a range of economic and social indicators and to appraise the state of its infrastructure. The main aim is to analyse the demographic and economic evolution undergone by the *comarca* in the last twenty years and to evaluate the present-day situation from the position of the duality, the differences, existing between the town of Solsona and the rest of the territory. The indicators used should enable us to confirm the initial hypothesis that this zone of Catalonia can hardly be considered a marginal area today (see e.g., Sánchez-Aguilera, 1996).
- To analyse the varying perceptions of the marginality of a given rural space due to the changes in values and behaviour brought about in a predominantly urban, western society, in relation to the rural space. Currently the rural world is undergoing a process of revaluation, now seen as an area that is full of positive connotations giving rise among urban populations to a growing interest, anxious to know the reality of these territories. A clear example of this is the appearance and consolidation of rural tourism. In recent years, thanks to these new tendencies as to how rural areas are perceived by a part of society, we might conclude that there has been a reduction although not a complete disappearance of the perceived marginality, which traditionally was associated with rural areas.

The reality and the perception of a rural space in the interior of Catalonia

The subject is wide-ranging and complex and, therefore, we must seek to briefly summarise those aspects, which according to the case under analysis we feel to be most relevant and meaningful of this dual interpretation of *comarcal* marginality. It should not be forgotten that we are analysing a rural space which in the last 20 years has undergone major transformations as a result of being subjected to constant change, a process which has been accelerated in the last decade. Indeed, it is possible to classify it as a dynamic space in a social and economic context (Aldomà and Pujades, 1987).

A *peripheral comarca in the geographical heart of Catalonia*

Today, and this despite the improvement in the road network, the *comarca* of Solsonès remains relatively isolated geographically from the main centres of consumption and the market. For centuries the area has suffered extreme isolation and this has been one of the facts that has contributed most to defining the nature of its landscape and cultural features which survive to the present day. We are speaking of a physical isolation since the zone is far removed from the large cities and main consumer centres (Barcelona, 115 km., Lleida, 110 km.), and what is more an isolation that is exacerbated by the inadequate nature of the road transport network in the area. For this reason, Solsonès may be considered a 'peripheral' *comarca* in socio-economic terms, in spite of being in the geographical centre of Catalonia. The absence of rapid transport links connecting the *comarca* with the main exterior market centres is a factor which *a priori* impedes the installation of new industries and services and, at the same time, highlights the isolation which we mentioned earlier (Soy, 1995). It is the relative isolation of the zone, which we consider the starting point for this article in order to explain the socio-economic characteristics of the *comarca* of Solsonès.

The *comarca* of Solsonès is situated between the Prepyrenees and the Central Catalan high plateau, occupying a set of high altitude lands, which correspond, to the inter-fluvial zone between the basins of the Llobregat and the Segre. It is these characteristics of the physical environment that have helped shape the demographic and economic characteristics of the zone, since it has always been a highly inaccessible territory. The main communication channels, carrying the largest flow of people and goods have always tended to follow the course of the rivers. In the river valleys and the plain areas is where most of the population is concentrated. Therefore, the *comarca* has traditionally been inadequately linked with the exterior and it has remained on the margin of the main communication channels: the axis formed by the river Llobregat and the river Segre. In recent years, attempts have been made at strengthening the axis of the river Cardener as it represents the shortest distance between Barcelona and Andorra cutting across the *comarca* from east to west. The improvement and strengthening of this access road will benefit the connectivity of the *comarca* with the metropolitan area of Barcelona, which has always been its natural outlet to the sea. Thus, the territory given its orography can be considered marginal, yet due to the improvement of the communication channels and the increased accessibility; today it no longer makes sense to call Solsonès a marginal territory.

Certain human aspects, if taken in conjunction with certain of the aspects touched on above, might conspire to ensure that the zone is still marginal. First, human settlement is in the main dispersed with the exception of the town of Solsona and the village of Sant Llorenç de Morunys. The dispersed nature of the population lowers the level of connection between the inhabitants of a particular zone and, what is more, is an additional difficulty to that of the relief contributing to make good internal communications a very expensive proposition. Due to these circumstances a large part of the subsidies and grants that the *comarca* receives from the European Union (it qualifies as an Objective 5b territory) are aimed at improving the road infrastructure of the municipalities (subdivision of the *comarca*), despite the enormity of the task given the large number of *masias* (farms) scattered over the area. Second, population density is low, less than five inhabitants per square kilometre in most of the municipalities, with the exception of Solsona where more than 60% of the *comaraca's* population is concentrated. A priori, and unlike the high Pyrenees, most of the territory is inhabitable, since the mean altitude of the zone, which stands at between 700 and 900 m, and the slopes that are not very steep favour the agrarian practices throughout the territory and yet there are wide areas which are poorly populated. In this the highest zone of the interior of Catalonia, population density is very low which links it with the mountain areas of the Pyrenees and the Prepyrenees. The lack of demographic power of the zone and the dispersed settlement pattern will have repercussions on the precariousness of infrastructure and service provisions.

The duality Solsona-Comarca and the marginality of a territory

As in other rural areas, in which the population densities are relatively low, the population tends to concentrate in certain points where most of the basic facilities and services are located. Due to the crisis in the agrarian sector, the processes of depopulation and the reduction in the number of active farms in most of the *municipios* continue. Meanwhile the population tends to become concentrated in those nuclei which have the administrative function of being heads of the *comarca*, the market centres of the surrounding rural spaces. At the scale of the *comarca*, this phenomenon has become more marked since the crisis began in the mountain zones, in the agrarian sector. This is a general trend in most of the rural *comarcas* in the interior of Catalonia. In the case of Solsonès this trend is particularly apparent.

This duality in the processes between the capital and its *comarca* is more evident than ever these days since Solsona has become the economic and social motor for its area of influence and is responsible for making Solsonès a dynamic *comarca*. The influence of the capital over the surrounding territory is such that in recent years the demographic (increase in population) and economic trends are visible in the general indices of the *comarca* and its power of attraction, as a market centre, goes well beyond the strict limits of the *comarca*.

In the demographic evolution experienced by the town of Solsona, over the last 20 years, more than evident are the direct consequences on the population of the function undertaken by the town as the capital and service centre of the *comarca* (attraction of the population). Recently, the population of Solsona has continued to grow, albeit moderately, and from a population of some 6,000 inhabitants in 1975 it has grown to more than 7,000 in 1996 (7,128 inhabitants). However, in the rest of the *comarca* the evolution of the population has shown a gradual fall. In the last five years, the increase in the number of inhabitants in the city of Solsona has been large enough to affect the overall trend of the *comarca* giving a positive growth rate to the population. However, it should be noted that most of the population increase in Solsona is due to migrations from other places in the same *comarca* and it is the main receiver nucleus of recent arrivals, in the main from the same *comarca* or those neighbouring it.

This divergence in trends is also evident in economic terms and in the characteristics of the job market. On the one hand, around 70% of the work force in the smaller and least populated municipalities – on average in those municipalities with less than 500 inhabitants – is primarily engaged in the primary sector. In contrast, in Solsona these percentages are reversed, as here it is the industrial and service sectors in which the greatest proportion of the population is engaged. Here also the *comarcal* statistics reflect the trends within the town of Solsona: in 1991 21% of the population of Solsonès was engaged in the primary sector, 29% in the industrial sector, 14% in construction and 36% in the service sector.

Other socio-economic indicators

The inhabitants of the *comarca* of Solsonès enjoy living standards that are equal to or better than those of many citizens of Catalonia, if we measure these in terms of social and job indicators and the income of its inhabitants. The unemployment rate in Solsonès is well below the average for Catalonia. In December, 1996 unemployment stood at 5% while in Catalonia the unemployment affected 10% of the active population. The *comarca* is not unique in this respect as most of the rural *comarcas* in the interior of Catalonia record unemployment rates well below the regional average. A further indicator is the gross disposable family income per inhabitant, which between 1986 and 1995 increased gradually attaining a value very near that of the average for Catalonia. In 1995 this income per inhabitant in the Solsonès stood at 1,412.8 thousand pesetas, while in Catalonia the average was 1,458 thousand pesetas. A final indicator is that of car ownership which in Solsonès is among the highest in Catalonia in relation to the number of inhabitants. In 1996, there were 453.6 cars per thousand inhabitants. In part, this high rate is explained by the dispersed nature of the population and the lack of public transport, which makes car ownership almost essential.

The provision of infrastructure and services in the *comarca* is affected by the problems that have their root in a scattered, disperse population and, therefore, this hinders the provision of all basic services to the population in its entirety, since the installation costs incurred are very high. Thus, there were only 41.6 telephone lines in operation per 100 inhabitants in 1996, while in Catalonia there were 46.7. However, while this service is still inadequate the last five years have seen considerable improvements. Thus, if we compare these figures with those from the year 1992 a rapid improvement can be seen in the coverage of telephone services in the zone - in Solsonès the figure was 34.1 telephone lines for every 100 inhabitants while in Catalonia the figure was 43.7. Second, the provision of health care in the *comarca* is particularly precarious, above all the number of available hospital beds per 1,000 inhabitants. In 1996 there were 1.07 beds per 1,000 inhabitants compared to 5.38 in Catalonia as a whole. These shortcomings in health care facilities and infrastructure are still evident and Solsonès is the *comarca* in Catalonia with the fewest hospital beds. In short, in the provision of health care, the *comarca* of Solsonès has a long way to go before it catches up with the average levels of Catalonia.

The perception of Solsonès: the comarca seen from outside its borders

Until a few years ago, certainly no more than a decade, there were a series of elements, such as the demographic insignificance of Solsonès within Catalonia, the fact that no major road passed through the area, the view of the rural world held by the inhabitants of the urban areas, which favoured a general ignorance of the *comarca* of Solsonès among the rest of the inhabitants of Catalonia. This ignorance of the territory can be correlated with the marginal location occupied by this territory in the perception of the people living outside it as there was no specific fact that awoke their interest.

The perceived marginality of Solsonès, due to this lack of knowledge, situates the *comarca* in the mental map as being much further from the metropolitan area of Barcelona than it really is. Recently, to a greater or lesser extent, within Catalonia the whole of the region has experienced a rise in the mobility of its inhabitants, in part due to the substantial improvement in the communication network. Most of the territory is now highly accessible and this has had the effect of making those regions that traditionally were most isolated from the urban centres more dynamic. In the context of these general processes, Solsonès is beginning to emerge from its state of anonymity, a condition bestowed upon it by the majority of Catalans until only a few years ago.

The post-productivism of the rural areas is due, in part, to the new tendencies in the perceptions of the countryside and rural areas in general of a large part of the population. The countryside is losing the pejorative connotations of being a backward area both economically and culturally. A revaluation of this space is occurring thanks to the increased interest in the understanding of those aspects which go to make up the area and one of the ways in which this tendency is shown is by entering into direct contact with this space through rural tourism (López-Palomeque, 1996). The rural tourism phenomenon, while diversifying the economic activity of the zone, also reflects a change in the mentality of a part of the society. Increasingly, aspects related to the environment are more highly valued as people rediscover nature and, in general, those elements that are not present in their daily lives (peace, nature, etc.). There is an interest in discovering the elements which form part of the traditional rural world and based on this the countryside and the cities, the rural world and the urban world have enjoyed a rapprochement, above all in cultural aspects, and the traditional image which associated rural areas with marginal areas has been eroded. In the case of Solsonès, rural tourism has contributed, in part, to the promotion of the *comarca* and in the diffusion of its reality to those living in large urban agglomerations with a very distinct daily existence. Thus, the marginal perception is disappearing even though many of the stereotypical images prevail in the mentality of many people who continue to view rural areas negatively - the fruit of their own ignorance.

Looking elsewhere and as often happens in those places about which little is known and which are far from the main centres of population and information, these zones frequently go unnoticed in our daily lives unless an event of great magnitude, usually a catastrophe, puts them in the headlines of all the mass media. In the case of Solsonès, last July a large forest fire devastated more than 15,000 ha of woodland and for a week the *comarca* was on the front page of all the newspapers. Unfortunately, these events might provoke new socio-economic problems, which will slow down the process of economic dynamism within the *comarca* yet, on the other hand, such catastrophes, in terms of perception, put the *comarca* on the map of Catalonia.

Conclusions

The socio-economic tendencies and processes of the rural areas of the interior of Catalonia have led to a reduction in the marginal character of most of these spaces, in relation to the most dynamic and prosperous areas of Catalonia. Today it is difficult to identify territorial marginality in Catalonia although a number of small isolated enclaves might survive, in the most inaccessible zones of upland areas. In contrast, what we are witnessing is a growing marginality of a social or cultural kind derived from the general processes generated within the market economy system and western society and in the era of the telecommunications revolution.

In the *comarca* of Solsonès, the current trend in demographic and economic processes continues to suggest a growing divergence between the capital and the rest of the *comarca*. The power exercised by the capital on the *comarca* is translated at the *comarcal* level in demographic and economic growth and increasing income levels, so that, without any doubt, the town of Solsona can be considered as the economic and social motor of the surrounding territory. Thus, the *comarca* of Solsonès can no longer be considered an isolated area, and, as such, marginal to the rest of Catalonia.

A change is occurring in the perception of the rural space. A growing interest is being expressed in discovering the countryside and this phenomenon is leading to the gradual disappearance of the pejorative connotations traditionally associated with these areas (Table 12.1). Thus, we can conclude that, at the level of perception, a large part of the urban population is today fully conscious of the better quality of life enjoyed by the inhabitants living in rural areas near to nature. In spite of this there remains a part of society who continues to ignore rural areas and, therefore, in terms of the perception of the interior of Catalonia it finds itself in a phase in which, on the basis of certain elements, it might still be considered as marginal (see Table 12.1).

Table 12.1 summarises the main characteristics, which might be used in defining the *comarca* of Solsonès as a marginal area or otherwise in relation to the region of Catalonia. The following points seek to follow the evolution over the last twenty years.

Table 12.1   Real and perceived characteristics of the *comarca* of Solsonès

REALITY	PERCEPTION
ASPECTS DEFINING THE DISTRICT AS MARGINAL	
Isolated *comarca*	Ignorance of the *comarca*
✓ Far from major market centres	✓ Lack of demographic, political and economic strength with respect to the rest of Catalonia
✓ Inaccessible	
Dispersed population	✓ Lack of appreciation of the rural way of life among a sector of the population of Catalonia
✓ Ageing population	
✓ Predominance of the over 45 year old age group	
✓ Transport infrastructure in poor conditions	✓ Relatively isolated, reinforcing the perception of a *comarca* disconnected from major market centres
✓ Deficiencies in health, cultural and telecommunications infrastructure	
✓ Predominance of the agricultural sector as employer of the workforce	
✓ Depopulation of the area	

Table 12.1 continues ...

ASPECTS DEFINING THE DISTRICT AS NON-MARGINAL

Solsona (capital)

✓ Culturally and economically serves as engine for the *comarca*
✓ Demographic and economic growth
✓ Internal migration: a positive balance to the *comarca*

End of the dualistic way of life, city or district

✓ Appreciation of rural space and a growing interest in enjoying the environment
✓ Growth of 'rural tourism' as a promotional feature of the *comarca*

High standard of living

✓ Low level of unemployment
✓ Family income similar to the Catalan average
✓ High index of private cars per 1000 inhabitants
✓ Increase in commuting between *comarcas* for reasons of work. Influence of the Barcelona Metropolitan Area
✓ 'Cardener route' - importance of the Barcelona-Andorra route through Solsona

References

Aldomà, I. and R. Pujades (1987), *L'economia del Solsonès. Aprofitament integrat dels recursos comarcals*, Col. Comarques de Catalunya, Caixa de Catalunya, Barcelona.
Cors, M. (1998), *Turismes alternatius a l'interior de Catalunya: L'agroturisme i les residències-casa de pagès al Solsonès*, Tesi de Llicenciatura, Departament de Geografia Física i Anàlisi Geogràfica Regional, Universitat de Barcelona, Barcelona.
Font, J., Majoral, R. and Sánchez-Aguilera, D. (1997), 'Mobility of Marginal Areas in Catalonia. A case in the Pyrenees', in G. Jones and A. Morris (eds.), *Issues of environmental, economic and social stability in the development of Marginal Regions: Practices and Evaluation*, Departments of Geography, Universities of Strathclyde and Glasgow, pp.205-217.
Generalitat de Catalunya (1998), *Estadística comarcal i municipal, 1997*, Institut d'Estadística de Catalunya, Barcelona.
Generalitat de Catalunya (1998), *Cens de població, 1996*, Institut d'Estadística de Catalunya, Barcelona.

Jussila, H., Leimgruber, W. and Majoral R. (1998), *Perceptions of Marginality. Theoretical Issues and regional perceptions of marginality in geographical space*, Ashgate Publishers, Aldershot.

López-Palomeque, F. (1996), 'Rural Tourism as a strategy in the development of marginal areas. The case of Catalonia', in M.E. Furlani de Civit *et al.*, (eds.), *Development issues in Marginal Regions II. Policies and Strategies*, Universidad Nacional de Cuyo, Mendoza, pp.49-62.

López-Palomeque, F. and Cors M. (1998), 'Estratégias de diversificación de las explotaciones agrarias. Actividades de turismo alternativo en Sant Mateu de Bages (Cataluña)', *IX Coloquio de Geografia Rural*, AGE, Vitoria-Gazteiz, pp.131-139.

Majoral, R. and Sánchez-Aguilera, D. (1998), 'Remaining Marginal areas in rural Catalonia', in H. Jussila, R. Majoral and C.C. Mutambirwa, (eds.), *Marginality in Space -Past, Present and Future*, Ashgate Publishing Ltd., Aldershot.

Sánchez-Aguilera, D. (1996), 'Evaluating Marginality trough Demographic Indicators', in R.B. Singh and R. Majoral (eds.), *Development issues in Marginal Regions. Processes technological developments and societal reorganizations*, Oxford & IBH Publishing, Delhi, pp.133-148.

Soy, A. (1995), 'Estudis comarcals: Alt Urgell, Alta Ribagorça, Berguedà, Cerdanya, Pallars Jussà, Pallars Sobirà, Solsonès i Vall d'Aran', *Anuari Econòmic de Catalunya 1995*, Caixa de Catalunya, Barcelona, pp.247-270.

13 Territorial marginalization: Regional borders and globalization, some examples from Spain

HUGO CAPELLA AND JAUME FONT-GAROLERA

Frontiers in a Global World

Frontiers frequently constitute marginal zones. In some cases this is because the political frontiers have been established in sparsely populated and marginal territories (Vilar, 1980) that may be considered natural frontiers (deserts, marshlands, mountain ranges, large forests); while in others, it is because wars and treaties have carved up culturally homogenous spaces (Barth, 1969), with the resulting application of contrasting policies on both sides of the dividing line (Leimgruber, 1997). However, what both have in common is that many frontiers are or tend to become marginal zones.

In recent years, during a process of economic globalization (Ramonet, 1998), the existence of frontiers has become an obstacle to the creation of large internal markets such as that of the European Union (EU). This has meant that in the EU, in particular, interstate frontiers have tended to be suppressed, while at the same time, development policies have been put into effect in those frontier areas that had been marginalized for centuries (e.g. INTERREG).

This process, however, has given rise to two paradoxes: Firstly, it has accentuated regional differences and, as a consequence, strengthened the role of the borders between regions; and secondly, strengthened the external frontiers of the EU. Thus, within the EU the disappearance of the state borders has been achieved at the expense of reactivating regional boundaries and reasserting the differences with non-European territories. Thus, in this case, globalization, rather than breaking down the territorial marginalization, as was thought, has made it more acute, both on the global scale (external frontiers of the EU) and on the regional scale (inter-regional borders) opening up old divisions.

In this article we analyse three examples of this new situation (see Figure 13.1); first, the consequences of suppressing the Spanish interstate boundaries with Portugal (the area of Galicia and the north of Portugal) on the one hand, and France (the enclave of Llívia) on the other; second, the consolidation of the inter-regional boundaries following the regionalisation and decentralisation of Spain, examining the cases of the Condado de Treviño, in the Basque Country, and the Franja Aragonesa, on the frontier between Catalonia and Aragon; and finally the external frontiers of the EU in Spain, focusing on Ceuta, Melilla and the unique position of Gibraltar.

Figure 13.1 Location map

Different borders

Interstate frontiers in the EU

Spain's entry in the EEC in 1986, and the Schengen space in 1993, which consolidated the freedom of movement in 7 of the 15 member states of the EU (Spain, Portugal, France, the Benelux countries and Germany) meant the gradual official suppression of two of the oldest frontiers in Europe: those of Spain and Portugal and Spain and France.

The frontier between Spain and Portugal, around 1,000 km long, was laid down in the Treaty of Alcañices, signed in 1297, with the exception of the town of Olivença, which was ceded to Spain in the XVIII century. The consequences of the presence of this frontier that divided what were relatively homogenous territories for 600 years are more than apparent. Initially, the frontier strip was not very clearly defined. The two peoples living in the zone on either side of the frontier maintained many economic and cultural links and even took advantage of the distance from the two centres of power, in Madrid and Lisbon, and so enjoyed a certain autonomy. In Portugal, for example, the frontier, known as the *raia*, is considered to have its own distinct personality. The frontier - now seen as a relatively impermeable space - was strengthened during the 17th and 18th centuries, when both states reorganised their territories in a much more hierarchical structure governed from a centralised point. From this moment on, the effect of the frontier was to become more marked for two main reasons: First, because of Spanish ambitions to unify the Iberian Peninsula under her rule (17th-18th centuries); and secondly, because of the highly artificial nature of the frontier, which divided river basins and plains, and which could not be justified by any 'natural' criteria, having even been described as absurd (Campesino, 1994). Thus, the frontier zones became more and more separated, their backs turned to each other, as they acquired the dubious distinction of becoming the most marginal areas of their respective countries (in Spain, Las Hurdes between Salamanca and Extremadura and in Portugal the region of Trás os Montes). In the 20th century, the Portuguese frontier became the object of conflict, particularly with the construction of the large reservoirs on the frontier rivers of the Duero, Tajo and Guadiana, which led to tension in relations concerning the use of the water resource.

With the entry of both countries in the EU and the subsequent theoretical lifting of the frontiers, this wide frontier zone (la *raia*), has become the point of focus for various development plans (linked to the INTERREG Initiative of the EU), which are not exactly devoid of their own tensions and contradictions. On the one hand, in Portugal, the impact of the Spanish cities of the so-called *Ruta de la Plata* (Silver Route) on its territory is viewed with a certain reticence. The Spanish cities of Zamora, Salamanca, Cáceres and Badajoz are larger and more attractive as service centres than the small Portuguese provincial capitals of Guarda, Castelo Branco, Portalegre and Beja. Moreover, the central governments of both countries are also a little reticent to see the establishment of a Euroregion linking the north of Portugal (the Oporto area) with Galicia (Vigo).

The 450 km of the frontier between Spain and France lie largely in the Pyrenees. While considered a particularly old frontier, it was not established definitively until the 17th century, with the signing of the Treaty of the Pyrenees, between Spain and France (1659). Where possible, it was sought to ensure that the frontier coincided with the watersheds, although this criterion could not always be adhered to. Moreover, the existence of the Val d'Aran, Cerdanya, Andorra, and another 15 minor anomalies (Sermet, 1983) meant that this criterion was purely theoretical. It should perhaps also be borne in mind that the people of Navarra in the Basque Country and the Catalans occupy either side of the Pyrenees which are seen, therefore, more as a nucleus of concentration than as a dividing line.

One of the more unusual characteristics of the Pyrenean frontier is the enclave of Llívia (see Figure 13.2). It has an area of 12.8 km^2 and 924 inhabitants (1996) and is of Spanish sovereignty while entirely surrounded by French territory. The origin of this unusual situation can be traced to the interpretation of a clause in the Treaty of the Pyrenees. The Treaty ceded to France the Catalan counties – integrated in Spain – of Rosselló, Conflent, Vallespir, Capcir and part of Cerdanya (Anglada, 1962). In the case of the latter, the cession was stipulated of 33 villages, but as Llívia had the title of *Vila* (town), its cession was disputed and it was excluded from those territories ceded to France.

As an enclave Llívia became isolated and marginalized as it was forcefully cut off from the surrounding municipal districts and the law prohibited its urban and economic development. The urban nucleus, for example, lost its strategic position as a crossroads for the whole valley, on finding itself connected solely by one road to the rest of the Spanish territory. As a result of this the market centres were transferred to other places (Cutchet, 1974). However, it was unable to maintain - following much interstate litigation -

some traditional rights, such as the use of its summer pasture located in French territory. In spite of this, the people became increasingly aware of the enclosure in which they lived.

Figure 13.2   The enclave of Llívia

After 1986, following Spain's entry into the EU and the subsequent creation of the Schengen space (1993), various policies have been implemented - forming part of the INTERREG Initiative of the EU and other initiatives for cross-border co-operation - to re-establish the socio-economic and cultural links throughout the valley of the Cerdanya, with a particular concern for the situation of the municipal district of Llívia. In the case of the road network, for example, there are moves to re-establish the former network of roads which connected Llívia with the other neighbouring French villages. Also many joint policies are being implemented aimed at the promotion and development of tourism, particularly of ski resorts (Generalitat de Catalunya, 1995).

The suppression of the frontiers in the EU, illustrated here with the cases of the borders separating Spain from Portugal and from France, may spell the elimination of one of the main causes of territorial marginalization, apart from some cases, such as the urban centres that depended exclusively on their location on the frontier (For example, Portbou – see Font and Tort, 1996). However, most policies still focus on theoretical issues (strategic planning, promotion of co-operation), than on a real redevelopment of the frontier zones. Whatever the case, it has been noted that in those places where formerly cultural links existed the development and re-assimilation of the frontier zones is occurring much more rapidly (Galicia and Catalonia).

Inter-regional borders

Within the globalizing process of the European Union, the regions have acquired a basic role, following the decentralisation processes put into effect in several EU countries (e.g. Spain) and the application of the subsidiary principle (cession of powers to the government closest to the citizen: thus local government need not be the domain of regional or state governments). This explains why parallel to the gradual suppression of the interstate boundaries, strong regional powers have gained ground, with the result that the intrastate regional borders have gained in significance.

In Spain, the process of decentralisation was ushered in with the 1978 Constitution and the subsequent creation of the so-called state of Autonomies, which was a response to the demands of the historic nationalities (Galicia, the Basque Country, Navarra and Catalonia) and which later (after 1980) was extended to form the 17 Autonomous Communities or Spanish regions. With Spain's entry into the EU (1986), the process of regionalisation received indirect backing, as it coincided with the European policies of regional decentralisation.

In general, the historic Autonomous Communities have greater powers than the rest, though the situation is highly varied, in a constant state of change and, consequently, not always clear. Thus each Autonomous Community can establish its own legal frame of reference and specific powers in questions such as education, taxation, environment, health, police and linguistic policy. In this new legal framework, the frontiers or regional boundaries have acquired increasing importance, as each Autonomous Community seeks to affirm its rights over its territory, leading to various problems in the frontier zones where legal disputes over land go back centuries. Thus, Spanish territorial divisions include a large number of unusual situations, including enclaves.

This is the case of the Condado de Treviño (see Figure 13.3), a county of 222 km^2 and 1,107 inhabitants (1996) which administratively speaking belongs to the province of Burgos and the Autonomous Community of Castilla y León, but it forms an enclave completely subsumed within the Basque Country (province of Alava). This situation is due to the ties between the County and the kingdom of Castilla in the late 15th century. The county lies in the valley of the river Ayuda and comprises the municipal districts of Condado de Treviño (852 inhabitants) and La Puebla de Arganzón (265 inhabitants), which boast a large number of hamlets. In functional terms, the enclave of Treviño, far removed from the decision making centres of Castilla y León, remains a vital centre of communications between the Basque Country and the rest of the Peninsula, and has become a real obstacle for the development of the whole zone.

Figure 13.3   The enclave of the Condado de Treviño

Since its creation in the 15th century, seven attempts (1646, 1938, 1940, 1958, 1980, 1995 and 1998) have been made to integrate the Condado de Treviño within the province of Alava (Basque Country). At the last referendum held in the municipal district of Condado de Treviño (1998), 67.4% of voters were in favour of being incorporated within Alava and the Basque Country (Estavillo, 1998). It is thought that this result reflects functional reasons rather than political or cultural affinities. However, in spite of this, the situation remains unchanged because of the veto imposed by the regional government of Castilla y León (Gorospe, 1998) and the central government.

The failure to reach a definitive solution derives, in this case, from the fact that it has been turned into an affair of the state and used as an element in the discord between the Basque Country and Castilla y León in which the people of the County have virtually no say (hence the slogan: *¡Qué decida Treviño!* - 'Let Treviño decide!'). It would seem that the result of a local referendum is no longer sufficient, as the Condado de Treviño has become a symbol for the Junta of Castilla y León, a question of pride for the Basques and a potential minefield for the central government, fearful that the approval of this modification to the boundaries between regions might usher in an endless number of new conflicts in frontier territories (such as Bierzo between Castilla-León and Galicia) which find themselves in a similar situation.

A further case in which the regional boundaries have acquired renewed protagonism has arisen in the frontier zone between Catalonia and Aragón (known as the Franja de levante - the Eastern Strip - or poniente - the Western Strip, depending on whether one adopts the perspective of Aragón or Catalonia). It is governed by Aragón but is influenced culturally by Catalonia. Since the 12th century, for example, the bishopric of Lleida included land within Aragón, until quite recent dates (1998), when following a Papal decision, it was decided to adapt it to the prevailing administrative boundaries. Thus, the Vatican segregated 27 parishes - totalling some 65,000 inhabitants - from the diocese of Lleida and added them to the diocese of Barbastro-Monzón in Aragón, changing a state of affairs that had been maintained for more than 800 years (Echauz, 1998).

The consolidation of regional powers in Catalonia and Aragón has led, on the other hand, to a certain friction on the linguistic question and the recognition of the cultural diversity of this frontier zone. In any case, some parts of this area might suffer greater marginalization as a result of the establishment of real barriers by the regional governments. This creates a climate of tension in areas, which up until now had lived quite peacefully.

The external frontiers of the European Union

Spain, given its peripheral location, constitutes one of the external frontiers of the EU. While this frontier is largely a sea border, it also includes the enclaves of Ceuta (http:// www.ceuta.net) and Melilla in African territory

(Figures 13.4 and 13.5). The history of the evolution of the frontiers in these two territories might be defined as a process of growing separation between two worlds, the characteristics of which are increasingly different (Bravo, 1996 and Saro, 1996). The frontier is the symbol of the destruction of the Mediterranean melting pot which has gone from being the centre of a heterogeneous world to become the margin of an homogeneous, segregated, global world (Moscati et al., 1985). How in this case should we understand the globalization?

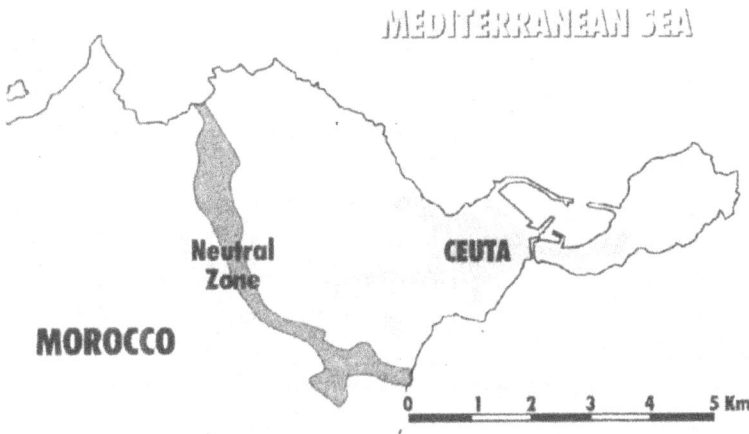

Figure 13.4   The enclave of Ceuta

Ceuta and Melilla are two Spanish enclaves in Morocco. They have an area of 19 and 14 km^2 and house 68,796 and 59,576 inhabitants respectively (1996). They are both densely urbanised and survive essentially on trade, defence and more recently tourism. Both of these territories have the unusual status of autonomous cities, i.e., the municipal council is also its autonomous government, and thus they comply with the Spanish policy of decentralisation. Each forms a part of the territory under Spanish sovereignty, and as such they are considered an integral part of the EU. However, and in spite of the fact that Spain forms a part of the Schengen space, Ceuta and Melilla have been excluded from the latter, so as to be able to apply the Law relating to Aliens more effectively and to control the constant flow of immigrants into the EU.

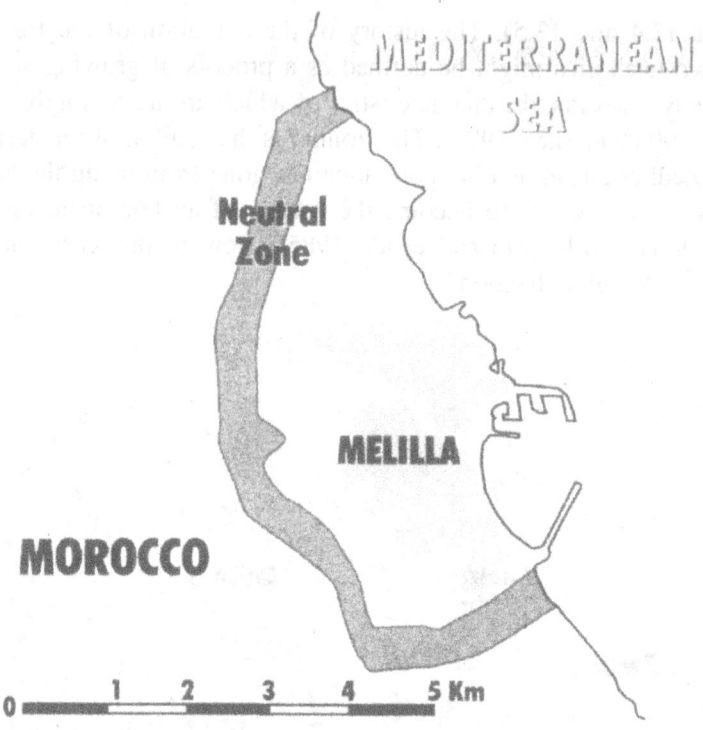

Figure 13.5 The enclave of Melilla

In both cases the frontier with Morocco, a perimeter of 9.5 km in Melilla and 6.5 km in Ceuta, is increasingly more difficult to penetrate, with a profusion of barbed wire, moats, cameras, police detectors and patrols. However, the immigrants who manage to cross these barriers find themselves on Spanish soil but not European as both cities are excluded from the Schengen space. They reach therefore an 'in-between' space - Ceuta or Melilla - which have been transformed into a kind of *no man's land*. The majority of the immigrants (some 2,000, on average) are crowded into the camps of Calmocarro and Ceutas awaiting authorisation to cross to the Peninsula. Normally, they refuse to state their place of origin so as to avoid being returned to Morocco. The latter, in its turn, refuses to recognise that the immigrants come from its territory or indeed that they have crossed it, as in this way Morocco can put pressure on the two cities, which it refuses to recognise as belonging to Spain.

However, both cities are considered territory of the European Union as far as the application of taxes and other aspects of EU legislation are concerned. Thus, one of the few incentives for these territories, namely exemption from taxes and custom duties, has disappeared leading to a slump in trade.

The Spanish government has taken into account the particular characteristics of these frontier cities and provides aid aimed at palliating their territorial marginalization. The main state and EU investments are usually aimed at strengthening the border perimeters, building refuges for immigrants as well as promoting conventional economic activities, which make up for the slump in trade, among others.

The territorial marginalization suffered by Ceuta and Melilla is comparable, in some aspects, to that of the Rock of Gibraltar, which occupies an area of around 5 km^2 and has a population of 30,000 inhabitants (see Figure 13.6) (http://www.gibraltar.gi). The situation of this territory, which has been under British sovereign rule since 1704, is one of the most complex and peculiar in the European Union. Gibraltar has, since the passing of its 1969 Constitution, been an autonomous territory, with virtually all its own powers (money, taxes, education, health) except foreign affairs and defence which continue under British rule. According to article 227 of the EC Treaty, Gibraltar now forms a part of the European Union, in virtue of being considered a European territory whose foreign relations depend on a Member State (Morris, D.S. and Haigh, 1992). However, in article 28 of the Treaty of Union (1972) a large number of tax exemptions were granted to the Rock, thereby turning it into a small tax haven (unlike its counterparts of Ceuta and Melilla).

The fact of belonging to the EU, but excluded from the space of free movement of goods, together with the situation that the United Kingdom does not form part of the Schengen space, has consolidated the frontier of Gibraltar with Spain. In fact, Spain closed the frontier between 1969 and 1975 and, since then, it has been the cause of constant conflicts. Today, the Spanish authorities continue to inspect merchandise, as Gibraltar does not form part of the Schengen space, although this affects, logically, the free movement of people. This results in endless queues (from 3 to 6 hour waits) causing serious bottlenecks in frontier traffic (annual movement totals some 14 million crossings in both directions). In spite of this, the frontier between Spain and Gibraltar does not benefit from EU initiatives for frontier zones (INTERREG).

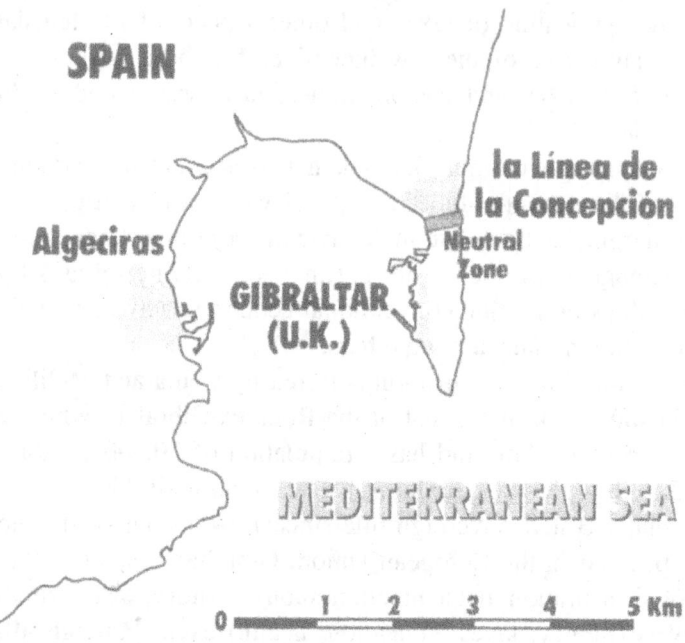

Figure 13.6   The Rock of Gibraltar

Territorial marginalization in a global world

Three conclusions may be drawn from this analysis of cases affecting Spain and the wider framework of the European Union regarding the role of frontiers in a global world:

a) The strengthening of the external frontiers of the EU

   The external frontiers of the EU, as illustrated by Ceuta and Melilla, seek to become ever more impermeable. This might seem to contradict the idea of hegemony perpetrated by globalization, given that it does not mean the automatic elimination of frontiers, nor the disappearance of social and territorial marginal areas to which they give rise. The notion of globalization is concerned with strategies for establishing financial and commercial order, rather than with universal social goals (conditions of work, the expansion of the welfare state, fundamental rights and environmental policies). It appears, in this case, that globalization has not acted in an effort to attain world cohesion, but rather it has brought about the internal unification of large markets, creating frontiers and divisions that constitute one of the main causes of inequality.

b) The progressive loss of importance, though with some variations, of the frontiers between the states of the EU

The formation of the EU should mean, in the medium term, the progressive lifting and disappearance of the frontiers between the states of the EU. This should alleviate the former territorial marginalization derived from the existence of frontiers, as was the case suffered by Llívia. It also favours, on the other hand, the re-establishment of ties and the integration of territories which were separated by the existence of a frontier, such as Galicia and the north of Portugal. On occasions, the disappearance of these barriers creates new situations of marginalization in towns and territories, which had previously depended for their living on the dividing line of the frontier.

c) The re-emergence of old regional borders

The growing importance of the regions and the affirmation of regional powers over a territory have led to the reappearance of former regional borders, resurrecting with them former tensions and marginal territories in detriment to the idea of globalization (the cases analysed of Condado de Treviño and the Franja in Aragón). This process is a further consequence of globalization. The establishment of the homogenising global model is developed more easily in a partitioned world (on a regional or local scale), where few possibilities exist for establishing bilateral contacts with ones neighbours. This spatial fragmentation and territorial marginalization are, consequently, just another strategy of globalization.

References

Anglada, M. (1962), *Vint i cinc anys a Llívia*, Editorial Selecta, Barcelona.
Barth, F. (1969), *Ethnicgroups and boundaries: The social organisation of culture difference*, Allen and Unwin, London.
Bravo, A. (1996), *La construcción de una ciudad europea en el contexto norteafricano*, Ciudad Autónoma, Málaga.
Campesino, A-J. (1994), 'Planificación estratégica transfronteriza en la raya lusa-extremeña', *Treballs de la Societat Catalana de Geografía*, 1994:37, pp.187-202.
Cutchet, S. (1974), *Llívia*, Editorial Altés, Barcelona.
Echauz, P. (1998), 'El obispado de Lleida consuma su reducción con la segregación de las últimas parroquias de la franja', *El País*, (14-06-98).
Echauz, P. (1998), 'El nuncio ordena devolver a Aragón el patrimonio artístico de la franja', *El País* (1-07-98).

Estavillo, D. (1978), *El Condado de Treviño*, Caja de Ahorros Municipal de Vitoria, Vitoria.
Font, J. and Tort, J. (1996), 'The study of a marginal area in the extreme east of the Pyrenees mountain range: the case of Port Bou', *Development issues in marginal regions*, Oxford and IBH Publishing Co. Pvt. Ltd, New Delhi.
Generalitat de Catalunya (1995), *Pla comarcal de muntanya de la Cerdanya*, Direcció general planificació i acció territorial, Barcelona.
Gorospe, P. (1998), 'Los treviñeses votan sí a la celebración de un referéndum sobre la segregación de Burgos' and 'Tres siglos con un pie en Burgos y el otro en Alava', *El País* (9-03-1998).
Leimgruber, W. (1997), 'Limites et frontières: thèmes permanents en Suisse. Le rôle des frontières dans nos réflexions et actions', *Treballs de la Societat Catalana de Geografia*, Barcelona.
Morris, D.S. and Haigh, R.H. (1992), *Britain, Spain and Gibraltar*, 1945-1990, Routledge, London.
Moscati, S. *et alter* (1985), *El mediterráneo otra vez; por una cultura del encuentro*, Editorial Encuentro, Madrid.
Ramonet, I. (1998), 'Firmes géantes, états nains', *Le Monde diplomatique*, p.1 (June).
Sermet, J. (1983), 'La frontière hispano-française des Pyrénées et les conditions de sa délimitation', *Pyrénées*, pp.11-37.
Saro, F. (1996), 'Notas sobre urbanismo, historia y sociedad en Melilla', *Estudios melillenses*, Melilla.
Vilar, P. (1980), 'El medi natural', *Catalunya dins l'Espanya moderna* (vol.-I), Edicions'62, Barcelona.

INTERNET sources

http://www.ceuta.net
http://www.gibraltar.gi

Acknowledgements

This paper has been prepared as part of the research project entitled *Delimitación y análisis de las áreas marginales en Cataluña*, funded by the Dirección General de Investigación Científica y Técnica (DGICYT) of the Ministerio de Educación y Cultura (Research Project: PB95-0905), and the (*Ajut de Suport a la Recerca dels Grups Consolidats del II Pla de Recerca de la Generalitat de Catalunya: Grup de Recerca d'Anàlisi Territorial i Desenvolupament Regional*, 1997SGR-00331).

14 Political boundaries and the development of marginal areas:
The border between Catalonia and Aragón (Spain)

JOAN TORT Í DONADA

Introduction

Among the host of internal borders within Spain (i.e. non-state borders), the boundary between the autonomous communities of Catalonia and Aragón is unique. It is one of the oldest (reputedly established in 1305, although its origins may well predate this by a number of centuries) and most firmly entrenched of all boundaries, given that it has not undergone any changes in the intervening centuries. Since first being drawn, the border has witnessed many significant events including the creation of the kingdom of Spain (1479), the administrative organisation of the state into provinces (1833) and subsequently into autonomous communities (1978). Yet, through all these upheavals the border has remained intact. Its uniqueness from a territorial perspective stems from the fact that its northern half (approximately 140 km) coincides with the course of one of the main rivers of the Spanish slopes of the Pyrenees: the river Noguera Ribagorzana (which flows into the river Segre and ultimately into the river Ebre). This would be of little significance if it were not for the fact that in most of the valleys of the southern slopes of the Pyrenees, it is the rivers that constitute the primary communication channels: the link between those living on either bank. The peoples of the valley form one community with strong, deeply rooted functional and social ties.

Here, the intention is not to offer a detailed analysis. Indeed the complexity of the issues requires a much more thorough analysis, which goes beyond the scope of this article. This article presents a geographical perspective based on observations carried out in the area, specifically the valley of the river Noguera Ribagorzana, paying special attention to the following:

a) The relationship between the valley and the rest of the Pyrenees;
b) The relationship between the origins of the border and the historical organisation of settlements;
c) The far-reaching crises suffered at the human, economic and social levels since the 1950s; and
d) The emergence of a sense of marginality among the valley's inhabitants due to the anomalous position of the border which runs counter to their everyday needs and interests.

The article concludes by drawing together the main aspects of the question in hand and considering briefly the future of the area.

The Noguera Ribagorzana valley and the Pyrenees

Lying in the southern slopes of the Pyrenees, the valley runs north to south – in line with most of the valleys in this range – and occupies an almost central position: some of its highest headwaters are practically equidistant between the Atlantic and the Mediterranean (Violant, 1949). Its basin, limited by the basins of Esera-Cinca to the west and Noguera Pallaresa to the east, is 140 km long if we follow the river axis, and 30 km across at its widest point. Its size – making it one of the largest rivers in the hydrographic system of the Pyrenees – and the marked contrast in altitude along its course – it falls from its highest point at 3000 m to 200 m – give rise to a wide variety of landscapes. In general, its upper course is typified by an alpine landscape with the bio-geography of high mountain areas; its middle course runs through the complex relief features of the pre-Pyrenees, marked by numerous gorges and small hollows; while its lower course flows through the wide sedimentary plain of Pla de Lleida (Solé, 1951).

However, it is the extraordinary degree of fragmentation within the valley, a feature it shares with other river basins in the Pyrenees of similar discharge and size, which prevents us from speaking of a valley in the true sense of the word and which gives the basin the appearance of a complex puzzle. The valley comprises a string of pieces constituting distinct units, physically very different and with little in common, at least historically. This is particularly evident in the middle of the basin – in its middle course – because of the predominance of the pre-Pyrenees, spanning a distance from north to south of almost 100 km.

These then are the main physical features of this valley and indeed of much of the Pyrenees. Given the general characteristics of this mountain range, in particular the east-west trend of its relief features, many of the rivers on the southern slopes present a similar typology to that of the Noguera Ribagorzana. Therefore we need to look at factors of a historical or human nature to explain its uniqueness and in particular its uniqueness as a frontier.

The river Noguera Ribagorzana as a boundary: the historical factors

We have been at pains to emphasise so far that all along the Noguera Ribagorzana river valley there exists a permanent mismatch between its geography and administrative expression. In order to understand the reasons that have brought about this situation, it is necessary to examine the area's history. Two factors above all require examination, both for the extent of their influence and the consequences they had in the settlement of the area. They both date back to the early Middle Ages (8th to 10th centuries) and unfolded in parallel fashion. On the one hand, there was what we might call the 'defensive factor' (or, to use the expression coined by the historian Pierre Vilar, the Pyrenees as refuge); while on the other hand there was the 'antagonism for territorial control', that is the fight to control new territories waged by the medieval counties of Ribagorza and Pallars, bringing the two into direct confrontation.

The first of these, the 'defensive factor', can be seen in the settlement pattern throughout all the *comarcas* of the Pyrenees. Its expression is quite simple, fulfilling the role that the Pyrenees as a mountain bastion has played throughout the centuries – in particular during the medieval period – as a natural refuge from invading forces. The vastness of this mountain area, its labyrinthine structure and the importance of the rivers that cut through it had contributed to the fact that fairly large groups of peoples – autochthonous or from other regions – settled here. At times of unrest, the advantages offered by the Pyrenees, in particular safety, were much greater than those offered by the wide open valley of the Ebre which ran some distance to the south. It is therefore perfectly logical that medieval settlements grew up near the headwaters and along the banks of the main rivers as they flowed through their middle course. Equally logical is the fact that centred around these early settlements, areas of great political and economic power grew up. The ancient counties of Sobrarbe and Ribagorza, in Aragón, and those of Pallars and Urgel, in Catalonia, are good examples of small politically autonomous and economically self-sufficient territories which at one time operated as states in their own right (Vilar, 1966).

Against this background, the second factor, 'antagonism for territorial control', came into play. The present-day boundary between Catalonia and Aragón, following the river Noguera Ribagorzana, cannot be understood without considering that this border was the result of a compromise in the confrontation between the counties of Ribagorza and Pallars to secure their respective territories. The boundary line, subject to constant revision since the 9th century, remains the same today as it was eventually agreed in 1305. It is not therefore a boundary reflecting local interests, but rather a frontier which was created with clear political interests in mind. Its objective is not to 'organise' a given territory, but rather to define the limits of two powers: initially those of Pallars and Ribagorza; and subsequently (what might be considered their successors), Catalonia and Aragón. Thus, from the moment of its creation the line followed by this frontier was totally unconnected with the geography of the Noguera Ribagorzana basin.

The modern era: from frontier to marginal zone

Historians agree that it was in the Middle Ages that the Pyrenees enjoyed its greatest splendour. The gradual loss of importance suffered by the old counties coinciding with the emergence of new socio-economic and political forces – namely Catalonia on the hand and Aragón on the other, gave rise however to a new political organisation, in which the centres of power lay far removed from the mountains and central valleys of the Pyrenees. The river Noguera Ribagorzana has remained throughout the centuries a frontier river, but the ancient *comarcas* of Pallars and Ribagorza, together with most of the *comarcas* in the Pyrenees, have gradually seen themselves reduced to peripheral areas, characterised, since the end of the Middle Ages, by a severe demographic and economic stagnation. Against this background of decline, Spain underwent an administrative reorganisation with the drawing up of provincial boundaries in 1833. In this area, this division served merely to confirm the ancient river boundary, constituting now the dividing line between the provinces of Huesca and Lleida. In the nineteenth century, Spain's respect for administrative history and tradition, despite its contradictions, apparently outweighed any concern for the development and revitalisation of the *comarcas* of the Pyrenees.

This situation of abandon and marginality to which the Noguera Ribagorzana valley was subject – along with most of the Pyrenees – became increasingly apparent from the second half of the 19th century onwards, due to the industrialisation of the coastal *comarcas* and the gradual shift of population to these areas from the rural peripheral zones. One of the main factors causing the marginalization of the valleys of the Pyrenees was the delay in the development of communication links, which in the main did not reach these areas until well into the present century.

When the Noguera Ribagorzana valley was finally integrated into what might be termed the modern 'productive system', it did so at a rather late date and in a somewhat abrupt manner. Once its potential for hydroelectric power had been realised, an ambitious plan was put into action to exploit the river and its tributaries. In a thirty-year period (1950-1980) a dozen dams and numerous channels and underground waterways were built in the river basin. No stretch of the river escaped exploitation, while the river's headwaters (given the steepness of their gradients) and its middle course were the most severely affected (Sánchez, 1991). In the latter case this was due to the labyrinthine structure of the relief that allowed large dams to be constructed, some of which are truly spectacular feats of engineering (notable examples are the dams of Escales, Canelles and Santa Anna). Intensive exploitation of the river Noguera Ribagorzana has resulted in the regulation of more than 90% of the water course: a figure which is not only a record in Spain but also throughout Europe (Vallès, 1949).

And yet, despite the great feats of engineering and the levels of productivity attained, we cannot ignore the effects on the valley's inhabitants. Thus while the hydroelectric programme has exploited a vast area, it has also entailed the acceptance of certain inconveniences which will colour its future. Firstly, the valley floor has been directly affected (a large part of the valley's 140 km). It was here that the best agricultural lands were to be found, but now only the few inhabited areas remain unaffected. Secondly, the intensive nature of the hydroelectric programme has had a perverse effect, above all in the long term: the reservoirs form a chain along the course of the river creating not only a physical but a psychological barrier. In some sections the reservoirs stretch over 40 km so that the distance by road from one side to the other can mean a journey of up to 100 km.

We should not lose sight of the fact, when determining the extent of the process of marginalization attributable to the hydroelectric programme, that we are dealing here with a territory deeply afflicted by demographic and socio-economic crises. The indices of depopulation, particularly in the river's middle course (i.e. the pre-Pyrenees), and above all between 1950 and 1980, are among the highest of all the Pyrenees. Some municipalities have been completely deserted, and the number of abandoned villages rises to several dozen, especially in the more mountainous areas that lie some distance from the river. With this in mind, and if we consider that the goal of the hydroelectric programme was to produce energy for other *comarcas*, we reach a simple conclusion: the result at the local scale has been a marked increase in its marginal character and its dependence on more developed regions.

Frontiers and a sense of marginality in the Noguera Ribagorzana valley

This analysis of the importance of the 'frontier factor' in the Noguera Ribagorzana valley reveals two sides to the problem: first, the inadequacy of the historical border between Catalonia and Aragón, given its eminently political origins; and second, the creation of a broad zone – on either side of the political boundary – suffering socio-economic and demographic marginalization, as a result of the neglect shown by official institutions throughout the modern era. Yet, independent of these two main circumstances, a number of secondary factors have combined to aggravate these initial problems.

First, there are those problems of an administrative nature caused by the border itself. The enacting of the 1978 Spanish Constitution and its subsequent implementation have established a decentralised model of territorial organisation, known as the 'State of the Autonomies'. This model has endowed the regional governments (in this case Catalonia and Aragón) with considerable importance, in particular, in the management of public services. In principle, the geographical proximity of the government bodies to the area being governed should be an advantage (in that an improvement in efficiency should be achieved); though in reality it has to be wondered whether this is the case of this valley in the Pyrenees cut in two by a border. Many examples can be cited of the absurdities to which this situation has given rise: in the same valley two different authorities work to provide a series of services that need to be managed in a unified way: waste management, fire services, education, health, environment and many other services. The paradoxical situation exists whereby the same orographic and bio-geographic system (the mountains of Montsec) has been declared a protected area and a national park on one side of the border, yet lacks any such legal protection on the other. These problems tend to be worse in practice, given the low level of economic resources dedicated by the regional authorities to the peripheral *comarcas*, compared with the central or more developed ones (Burgueño, 1995).

Second, are the problems generated by the fact that the inhabitants of the entire valley of Noguera Ribagorzana speak Catalan (the valley forms part of the frontier zone between Catalonia and Aragón in which Catalan has been spoken since time immemorial), but because of the frontier Catalan is only an official language in the eastern part of the valley (i.e. the half lying in Catalonia). Not surprisingly, an administrative anomaly of this nature, made worse by the passing years, has created a feeling of inferiority among a large number of the valley's inhabitants (Fábregas, 1971). A feeling which in many cases has become a 'marginality complex', to the extent that in addition to having to live in a peripheral area, the population suffer a permanent affront: that of seeing how the authorities deal with identical situations in an unequal manner.

Finally, although perhaps of secondary importance, another problem serves to illustrate the range of anomalies faced by the inhabitants of the valley. It is an issue which has received much coverage in the media: the problem of the religious border (or, more precisely, the 'border between ecclesiastical dioceses'). Since the Middle Ages the valley of Noguera Ribagorzana has fallen under the auspices of the Bishopric of Lleida. Indeed, the strict separation between the lay and religious spheres in Spain, at least in relation to territorial questions, meant that for centuries the limits of the bishopric could extend deep into Aragón, following a much more rational geographical boundary line than its counterpart. However, recently, the pressure exerted by various political groups and lobbies, ignorant of the real problems faced by this territory, has led the Vatican to redraw the boundary: today the religious and administrative borders coincide adding to the already existing disparities.

Borders: Obstacle or aid to development

French historian, Lucien Febvre claimed in a study published in 1925 that, 'At heart, every geographical problem is essentially one of limits' (Febvre, 1925). If we observe the frequent contradictions to which any border, at whatever level and in whichever country of the world, gives rise, it is difficult to disagree with Febvre's claim. The statement leads us to the conclusion that a frontier, a limit, constitutes something much more than a simple line in spatial geometry. Each border, by the simple fact of fulfilling this role, acquires a political, historical and symbolic dimension which frequently has no rational explanation. Geographers have traditionally interpreted borders as 'cultural' artifacts, but this interpretation has on very few occasions served to change the state of affairs. Each border has its own history, its own *raison d'être* perhaps, its own set of characteristics; and only by understanding each border can we begin to improve the organisation and management of the territory.

The border between Catalonia and Aragón in the valley of Noguera and Ribagorzana is a good example of this. Fruit of a political compromise, its survival over the centuries reflects the extent to which we might talk of the effects of 'historical determinism' as opposed to those of 'geographical determinism'. Thus, while in the modern age man has been able to overcome (to a certain degree) the limits imposed by nature through the wonders of technological innovation, he has been unable to change, in any logical or rational way, the results of historical inertia or the ineptitude of many politicians.

In 1933, Le Corbusier laid the foundations for a new means of understanding urban design by prioritising the needs and aspirations of man. When referring to the problem which frontiers often represent for a suitable territorial planning, he said 'no administrative limit can aspire to remain forever unchanged' (Le Corbusier, 1933). Perhaps now, at the end of the 20th century, the time has come for geographers to heed his words and set about changing the current state of affairs.

References

Burgueño, J. (1995), 'Franja de Ponent' in J. Font and J. Tort, *Les terres de parla catalana. Catalunya, València, Illes Balears, Andorra*, Editorial Thema, Barcelona.

Le Corbusier (1933), *Principios de urbanismo (La Carta de Atenas)*, Editorial Martín, Barcelona.

Fábregas, X. (1971), *Entre Catalunya i Aragó. Viatge per la frontera de la llengua*, Editorial Selecta, Barcelona.

Febvre, L. (1925), *La Tierra y la evolución humana*, Editorial Pirámide, Madrid.

Sánchez, L. (1991), *L'aventura hidroelèctrica de la Ribagorçana*, (ed. From the author), La Pobla de Segur.

Solé, L. (1951), *Los Pirineos. El medio y el hombre*, Editorial Martín, Barcelona.

Vallès, J. (1949), *La cuenca del Ribagorzana*, Enher, Barcelona.

Vilar, P. (1966), *Catalunya dins l'Espanya moderna* (4 vol.), Editorial 62, Barcelona.

Violant, R. (1949), *El Pirineo español*, Barcelona.

Acknowledgements

This paper has been prepared as part of the research project entitled *Delimitación y análisis de las áreas marginales en Cataluña*, funded by the Dirección General de Investigación Científica y Técnica (DGICYT) of the Ministerio de Educación y Cultura (Research Project: PB95-0905), and the (*Ajut de Suport a la Recerca dels Grups Consolidats del II Pla de Recerca de la Generalitat de Catalunya: Grup de Recerca d'Anàlisi Territorial i Desenvolupament Regional,* 1997SGR-00331).

15 Microperiphery in Mega-Cities

THOMAS BLOM

Introduction

Wealthy inner city centres and poor suburban ghettos exist irrespective of whether we study smaller urban areas in Sweden and Portugal or Megacities in the United States and Japan, it is becoming increasingly apparent that we have got a society where, no matter whether they are defined on economic, social, ethnic or demographic principles, different groups of people tend to gather in different housing estates without contact with one another. In some cases it is only a street that separates them. Are strictly segregated cities a reality that we must accept? Is a new form of urban periphery emerging that differs from the periphery we experience and attempt to combat in many rural areas?

The empirical material in the present study is drawn from New York City (NYC). The study of NYC is restricted to the twelve community districts, which form Manhattan. The variables I have use are: population changes over time, the proportion of inhabitants receiving some form of income support, the proportion of inhabitants under 18 and ethnic composition. The variables are quantitative in character in order that they may be used to distinguish materially peripheral islands in an urban agglomeration. With this as the point of departure, I discuss that the centre and periphery from an immaterial perspective. However, I do not claim that this is a comprehensive study, I simply want to discuss the issue on the basis of a number of examples. The study is thus not statistically generalisable but rather analytical generalisable.

Perspectives

Periphery is a concept that is often associated with the problems of rural areas in the form of high unemployment, a low level of formal education, high sick leave, low average income, poor public transport, rural postmen, shops only open between 10 a.m. and 4 p.m., closed schools and problems at the factory (Blom, 1997). Can we use these terms to describe the periphery and peripheral phenomena in urban areas as well?

Centre and periphery may be seen as interdependent concepts as it is impossible to define the periphery without defining the centre. It is thus difficult and probably not meaningful either to make a clear distinction between what is classified as central and peripheral respectively. The scale is fluid and depends on the points of reference that are used to define the standard. Is it not, in fact, more relevant to talk of several different types of periphery and even of different types of centrality?

In the public debate, centre and periphery are often implicitly understood as synonymous with what is perceived as positive and negative respectively. Regions defined as central are often used as a yardstick for regions that are defined as peripheral.

Geographical sparseness, in the sense of few inhabitants per square kilometre and a low level of service, is a typical feature of many peripheral regions. This is a well known and familiar definition but is it so simple? It is, I believe, important to introduce further dimensions into the debate, which highlight the difference between material and immaterial peripherality and that between material and immaterial centrality.

Material peripherality refers primarily to a physical sparseness of buildings and service functions. This type of peripherality is measurable and can thus be analysed with the aid of quantitative methods. Immaterial peripherality is rather what might be termed mental or emotional peripherality. This form of periphery is difficult to define in precise terms since every individual in their consciousness perceives centrality and periphery in their own way. Two individuals who are in exactly the same place at exactly the same time may have diametrically opposed perceptions of what is central and peripheral from an immaterial perspective.

Peripherality is not merely a matter of being geographically located outside a defined centre. In seemingly developed central urbanised regions there is often a form of microperipheral island, which in certain respects has a traditional peripheral character as regards, for instance, unemployment and education (Blom, 1996a). Typical of this island-structure is that in some ways its character is similar to that generally found in rural areas, which are defined as peripheral. Perceived in mental or emotional terms, however, these microperipheral islands in urban agglomerations are experienced, as significantly more peripheral than is the case with many rural areas. The quality of life and what is perceived as welfare are significant components, which are often neglected in the traditional centre-periphery debate (Blom, 1996a).

Thus it is important to introduce a new dimension into the periphery debate and, in contrast to the more generally accepted approach, to base the discussion on the regional structure of urban agglomerations. My intention is to highlight the microperipheral islands, which are increasingly becoming a feature of today's urban regions throughout the world.

It is possible to make a distinction between this new dimension and the concept of segregation, which is used to separate one or more categories of people from the rest of the population, either as a step in a conscious policy or because of other circumstances which unconsciously lead to this separation. The point of departure and the significance to the concept of microperiphery has been attached a deeper meaning than segregation. The emphasis is on the individual's immaterial and emotional view of what is perceived as central or peripheral. Segregation is, however, a major and fundamental feature of the argument as it expresses a form of peripherality in contrast to what many have defined as central and more important. Thus I would maintain that peripherality is not something that can simply be defined in terms of geographical distance and material factors. Islands emerge in urban regions where the inhabitants experience both material and immaterial periphery, at the same time as inhabitants in rural areas can experience that they are in a material and immaterial centre.

The various dimensions of the concept of microperiphery

In many of the world's central urbanised regions a number of microperipheral islands are becoming increasingly apparent, which may in several respects be significantly more peripheral, primarily from an immaterial standpoint, than many rural areas. This microperipherality finds expression in such things as social exclusion, drug abuse, and apathy in young people who have no chance of finding a job, resulting in increased criminality and resignation. A further characteristic feature of the young people is that they have a relatively low level of education. These microperipheral islands are often thought to deviate in various ways from the normal structure of society and to be frequently populated by immigrants who have been further isolated from the rest of society through conscious or unconscious segregation. A new island culture is emerging which has little in common with the surrounding urban area. The image of the big city as the source and stronghold of the good life and economic growth is no longer undisputed as structures evolve in urban regions which in many ways are an expression of a peripherality problem which requires new solutions and new ideas (Blom, 1996b).

On the basis of studies of the 'Greater Detroit Area', Mehretu *et al.* (1996) show there are major intraregional differences in the area in that it is strictly divided along various lines into sectors where poverty, level of education, unemployment and ethnic structure produce a more or less pronounced polarisation. Studies of intraregional conditions and development problems have also been made by Hayter and Barnes, (1992), Law and Woch, (1993) and Knox, (1994). There are discussions of studies of segregation problems in McDowell, (1995) and Cadwallader, (1996) who raise the issue of the increasing 'social distance' in modern Mega-cities, with the resultant emergence of separate social zones within the city. The emphasis in the research that has so far been conducted has been on income, education and employment structures. The quantitative material aspects have been highlighted whilst the immaterial perspective has only been of secondary importance in the discussion.

In its final report, Storstadsutredningen (SOU 1997:118) (the Swedish government's study of big cities in Sweden) states that segregation in big cities in Sweden had grown rapidly during the 1980s. The survey shows that immigrants were over-represented in the 'least attractive' housing areas in the big cities, as were the unemployed and people who were more or less permanently excluded from the labour market. In these areas many households were dependent on income support. The proportion of households of this type was increasing at an above-average rate for the region. The proportion of early retirees, single-person households and single mothers was high. There was also a considerable overrepresentation of people in poor health. Many were on the sick list. The consumption of health care, and mortality were higher than average in the region. What has happened since then? The big city committee conducted a follow up study on the period 1985-1995. The situation must be considered serious as economic, social and ethnic segregation has been consolidated. The traditional picture of purely ethnic segregation is no longer valid; rather a clear economic and social segregation has evolved.

Different ethnic groups are often believed to have chosen to live segregated and not to want or to have been able to establish contacts with the majority of the native population (as a result, for example, of language difficulties). Those who were born in the country and are socially excluded are sometimes faced with another dilemma, as they hardly seem to exist at all for the majority of the native population. The link between alienation and economic conditions becomes very obvious (SOU 1997:118).

Peripherality in the usual sense of geographical distance from what is defined as central in a country or region is one side of the problem. Let us now consider the situation from a different perspective and examine peripheral phenomena in central regions. I want to highlight the problem of marginalization and peripherality, which is not only becoming increasingly significant in many of the big cities of the world but is also a growing problem in smaller urban areas, and which is more immaterial in nature.

A typical feature of big cities is that despite the fact that there are a large number of people in a limited space, few know each other personally. Contacts between individuals are perceived as superficial and fragmentary, an instance of immaterial microperipherality where ethnocentrism creates a threat and raises barriers.

In many of the microperipheral islands that arise in urban agglomerations there is an immaterial peripherality, which has to be dealt with by other means than the material peripherality facing many of the rural areas. Different microperipheral islands in one and the same urban agglomeration may, when they are compared, reveal diametrically opposed problems, each of which makes them peripheral but for different reasons – material/immaterial. The lack of social relations between different groups of people is, however, an increasingly common variable for peripherality focusing on a clear unequal distribution of people – socially, economically, ethnically and demographically. This lack of contact results in a distance between various groups, which is manifested in their physical separation.

A relatively simple approach is to use income and educational levels as a measure of microperipherality. People are still classified today in accordance with a yardstick proposed by Charles Booth in 1889 (Giddens, 1994) where the object was to attempt to establish a viable measure for a subsistence level in order to define poverty. The point of departure was exclusively economic conditions. No attention was given to the general level of welfare or to what the individual defined as quality of life. If we introduce aspects such as quality of life and welfare into the discussion, it becomes much more difficult or perhaps even impossible to talk about rich and poor. Similarly, whether we live in a district that is defined as rich or one that is considered poor, there is nothing to say that the feeling we have for our home district or the desire to go on living there is different, irrespective of whether we live in a high-status area like Mayfair in London or a low-status area like the Bronx in New York. The problem lies more in the relation of the islands to each other and the feeling of belonging or alienation which exists in them and between them. There is a considerable risk in talking in general terms of the big city as a homogenous unit and a similar risk in homogenising rural areas. It is instead a matter of more or less obvious transitions, which change over time, depending in part on the life mode structure of the individual. The inhabitants of one city district may experience both material and immaterial centrality whilst those in another consider that they live on both a material and an immaterial periphery. Further, there are within a district considerable differences between individuals with respect to the life they live and strive for, since what is considered welfare and quality of life differs significantly from person to person. A Swedish family living in an area with many immigrants, where the majority of the inhabitants have a similar ethnic background may experience a form of immaterial peripherality which is probably comparable to that experienced by immigrants living in districts with a high proportion of Swedes.

Even if the conditions for material centrality in the form of infrastructure, supply of goods and services, schools, housing etc. are met in a given region, individuals may still experience material peripherality if they do not have the opportunity of enjoying this supply. Being unemployed, not longer receiving unemployment benefits, not having education or being ill in a society where personal economic resources are crucial is most probably perceived as tangible material peripherality despite the geographical proximity to the supply.

New York City – a Mega-City with peripheral ingredients?

New York City (NYC), situated on the estuary of the Hudson river as it flows into the Atlantic, is divided into five administrative boroughs: Manhattan, the Bronx, Brooklyn, Queens and Staten Island (Community District Needs, Manhattan, 1996). The Bronx is the only one of the five boroughs on the mainland.

In 1950 NYC was the largest urban agglomeration in the world with 12.3 million inhabitants. It is estimated that it will have 16.8 million in 2000 (World Urbanization Prospect, 1991). However, this refers to the so-called 'metropolitan area', which also includes Long Island, northern New Jersey and certain parts of Connecticut. NYC covers 837 km^2 and in 1990 had 7.3 million inhabitants. Brooklyn is the largest borough population-wise with 2,300,664 inhabitants in 1990 and Staten Island the smallest with 378,977 (Community District Needs, Manhattan, 1996), see Figure 15.1.

Since 1970 the total number of inhabitants in all five boroughs of the city has decreased by just over 7% or 573,336 individuals. If we make a closer study of the regional changes in population for each of the boroughs over the same period, we find that the population of Staten Island has increased by over 28% whilst that of the Bronx has decreased by over 18% (Community District Needs, Manhattan, 1996) (Figure 15.2). In fact, Staten Island is the only borough to show an increase in population in the period 1970-1990.

One of the major problems in a city like NYC is not primarily that the city is perceived as mentally tough and that the pulse is high but concerns housing. The rents in Manhattan are higher that anywhere else in the US. Those who are not well off or have official housing have to live either in

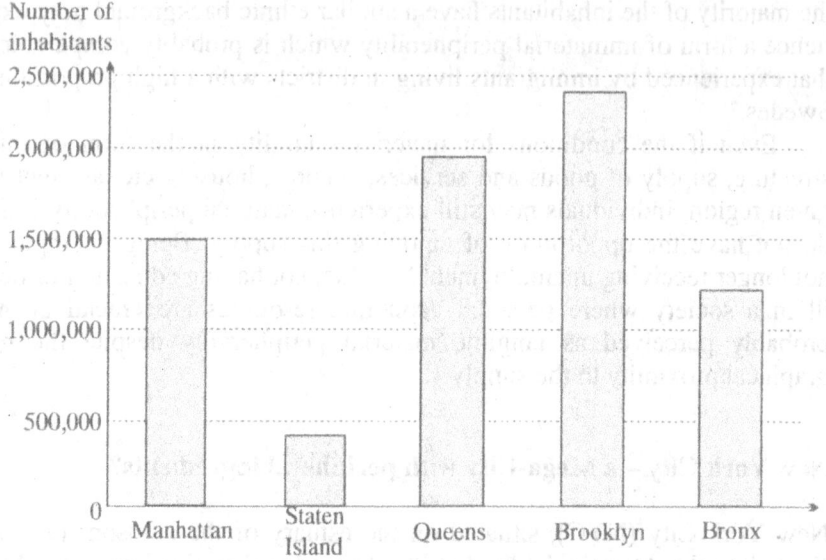

Figure 15.1 The number of inhabitants in New York City's five boroughs

Source: Community District Needs, Manhattan 1996.

poor housing or far from the centre in one of the gigantic dormitory suburbs surrounding Manhattan. These are either residential districts, highrise apartment blocks or slum areas. It is typical of NYC, as of many other cities in the US and in the rest of the world, that segregation in housing is more or less explicit. It is generally well known, at least among the city's inhabitants, which areas are occupied by the so-called upper class, upper middle class, middle class and lower class. In some environments terms like slum and ghetto are used to underline the negative status of certain areas and to distinguish them from the upper and upper middle class areas. In the same way, reserve of the rich is used in a derogatory sense for upper class areas in cities. The status of an area changes over time as, for instance, property speculators buy up 'slum' houses or whole blocks and completely renovate them. The old tenants are bought out or are forced to move and new, wealthier tenants move in (Cater and Jones, 1989).

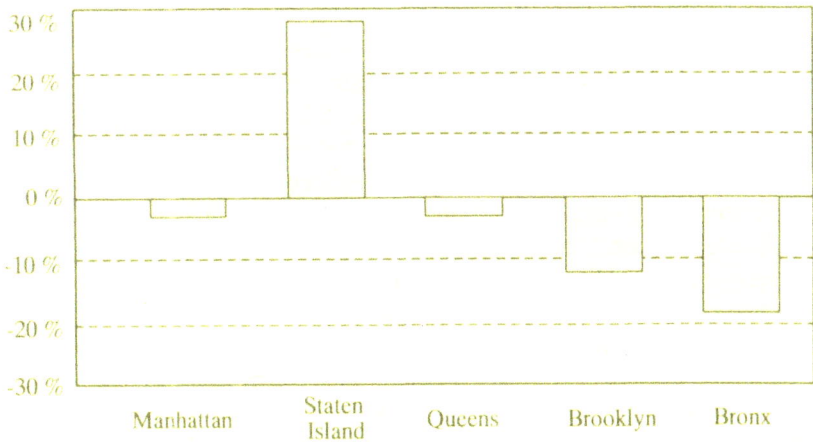

Figure 15.2 Changes in the population of New York City's five boroughs 1970-1990

Source: Community District Needs, Manhattan 1996.

In 1993 24% of the population of NYC were in one way or another dependent on assistance from the community. 14.8% received Public Assistance (AFDC, Home Relief), 4.3% Supplemental Security Income and 4.9% Medicaid (Cater and Jones, 1989).

For a long time NYC has been a strongly segregated city with major social problems. These differences are both of a regional and ethnic nature. The differences between the black and white populations are apparent in various ways. Unemployment among young blacks has remained constant at about 30% since 1972. The corresponding figure for young whites is 12% (Wille, 1993).

Living conditions have deteriorated drastically in recent years. One crucial difference between the situation today and earlier is that the poor generally used to live in a socially functioning context. They had relatives, the church, meeting halls and unions. Thus ethnic concentrations in certain parts of a city are not necessarily a negative factor in them as the proximity to fellow countrymen can create security and stability. Today, however, increasing numbers of people live in immaterial isolation far from work and a normal social life. Alienation and resignation are common. In NYC, districts in the southern part of the Bronx, Brooklyn, central and eastern Harlem are often portrayed as areas where these problems are particularly in evidence.

The ethnic sectorisation in the city is striking. New arrivals often choose to live in existing ethnically homogenous districts which, looked upon from the outside, appear to be relatively isolated villages with their own infrastructures and cultures. Ethnic microperipheral islands emerge where the cultural differences between the different islands and between the islands and American society are conspicuous.

Examples of ethnic islands are to be found in Queens and Brooklyn, where large Greek (Astoria), Italian (Carroll Gardens), Colombian (Jackson Heights), Scandinavian (Gravesend), Korean and Indian colonies have become established and are, in their turn, in constant transformation. Just a few years ago Canal Street in south Manhattan had an Italian stamp now it is Chinese. Part of the Jewish Lower East Side has over time become Puerto Rican. Crown Heights in Brooklyn is black and no longer Jewish. More black people live in Bedford-Stuyvesant, East New York and Far Rockaway today than have ever lived in Harlem. New York is the largest Jewish city in the world with more Irish than in Dublin. Almost a quarter of its population is black. Up to the 1970s most of them came from the South. Today the biggest immigration of blacks is from the West Indies. The city has about 1.7 million Spanish-speaking inhabitants, mainly Puerto Ricans (Wille, 1993).

To summarise, it can be said of NYC that it embraces a whole series of paradoxes and contrasts. Behind its magnificent skyline with skyscrapers exuding power and wealth there is a New York consisting of districts with a high proportion of immigrants suffering from a lack of education, illiteracy and unemployment.

Borough of Manhattan

This article concentrates the microperipheral studies on the Borough of Manhattan. The island of Manhattan consists of 12 community districts. The total population was 1,487,536 in 1990. In 1993 22.7% of the population were dependent on some form of assistance from the community. 13.2% received Public Assistance (AFDC, Home Relief), 4.7% Supplemental Security Income and 4.8% Medicaid (Community District Needs, Manhattan, 1996). The population figures refer to 1970 and 1990. The percentages for those receiving some form of income support are from 1992. The details concerning the proportion of the population under 18 and the proportion classified as white non-Hispanic are from 1990.

Centre and Periphery in Manhattan

Is it possible and justified to talk of peripheral phenomena in one of the world's largest urban agglomerations? The concepts of centre and periphery must always occur in pairs since the periphery can only be defined in relation to the centre and vice versa. With this point of departure it is, of course, possible to speak of more peripheral and more central parts of Manhattan. However, it must be stressed that it is not the geographical aspect that it is of primary importance here. Material and immaterial peripherality and material and immaterial centrality are of more than just geographical interest in a city like New York.

A survey of the twelve districts of Manhattan reveals a clear picture of a segregated city centre where the differences in actual living conditions are significant. The statistics on which the study is based only provide the material side of what may be classified as central and peripheral respectively. On the basis of the variables which are used in the study it is, however, possible to form a picture of which areas many people would class as 'good', 'less good' and 'bad'. If we assume that we are going to move to a city which we do not know much about, it is relatively easy to form a mental social map on the basis of statistics on average income, formal levels of education, ethnic structure, people under 18, those on income support etc. This map often gives us an idea about which area we feel 'we would fit in to'. We do not just look for a place to live but an environment, which fits our own way of life.

Many of the stereotypical 'images' we bear within us are often based on this type of statistics reinforced by the information conveyed by the media and rumour. We seldom have 'inside information' from those living in the various areas but often base our notions on secondary information. In this context it of interest to refer to Buttimer (1981) who underlines the contrast between the way an 'insider' experiences a place and the conventional manner in which an 'outsider' describes places.

As is apparent from the figures for the twelve districts given in Figure 15.3, there are major regional differences with respect to the variables: proportion of inhabitants under 18, proportion of white inhabitants and proportion on some form of income support.

Those districts of Manhattan which show a high relative proportion of whites (white, non-Hispanic) also have a low relative proportion of people on income support and a low proportion of people under 18. This condition is particularly noticeable in districts 1,2,4,5,6,7 and 8 (Figure 15.4). These districts lie mainly in the southern and central areas of Manhattan. One exception, however, is district 3, whose ethnic composition and even buildings show many similarities with the northern districts (Harlem).

District 3 is Lower East Side and its population is made up of 32% with a Latin American background and 30% of Asiatic origin. Blacks represent just over eight per cent.

If we only include the variables: proportion of whites and proportion on income support, there is a clear dividing line between districts 1-8, with the exception of district 3, and districts 9-12 (Figure 15.4). Further, there is an obvious relation between proportion of whites and proportion on income support.

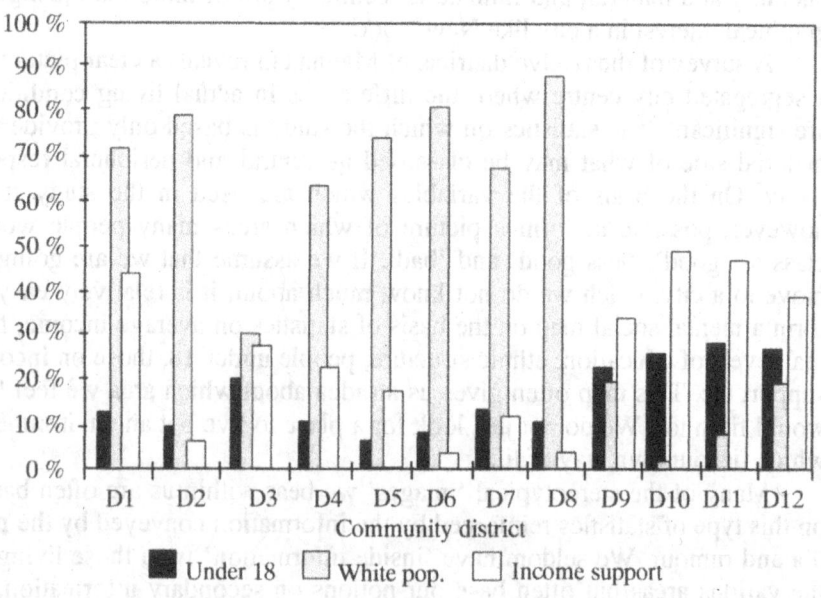

Figure 15.3 A comparison of Manhattan's twelve districts on the basis of three variables

Source: Community District Needs, Manhattan 1996.

Concluding reflections

Is it possible and relevant to speak of more peripheral and more central city areas on the basis of the variables used in this study? If we only apply a material physical perspective, It is possible to maintain that it is adequate to consider segregation between districts within a city. If, on the other hand, we wish to introduce an immaterial perspective, it is not sufficient merely to discuss segregation in the sense of being separated from the surrounding regions but to take into account a microperipherality which places emphasis on the individual's immaterial, mental and emotional view of what is perceived as central and peripheral respectively. In the public debate, however, reasoning on the status and character of different areas is generally based on a number of quantitatively measurable variables. On the basis of the results of these variables far-reaching conclusions are often drawn about the inner life and 'quality' of the area.

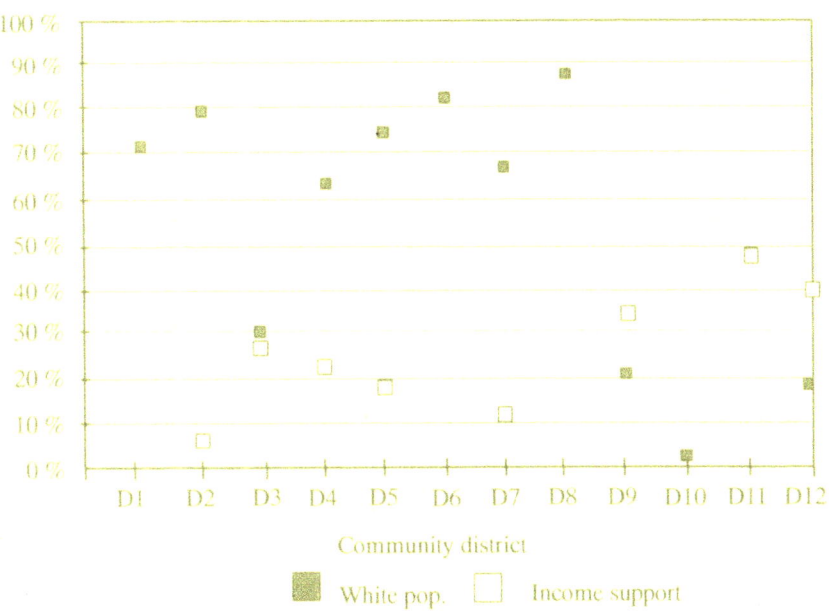

Figure 15.4 A comparison between the proportion of white population and the proportion receiving income support

Source: Community District Needs, Manhattan 1996.

There are indications that segregation in big cities is a matter of class differences between those who are established on the labour market and thus are also part of the welfare system and an *outsider* class.

The study of Manhattan and its twelve districts shows that it is possible to speak of segregation from a material standpoint and thus of a form of peripherality. If we instead take into account the individual's immaterial needs and the opportunities that provide them with the welfare and quality of life they desire, the picture of the twelve districts may well be very different. At the same time we may speak of class differences on the basis of quantitative variables and note that we have attained an increasingly refined class society in the form of, on the one hand, an 'outsider' class and, on the other, an *insider* class. The distinction is based on whether a person has a job. Getting a job, for instance, creates a social fellowship which can 'move' an individual from both a material and a perceived immaterial periphery to a material and immaterial centre, irrespective of whether the individual lives in a geographically defined periphery or in an urban agglomeration. Despite today's high technological and infrastructurally well developed society, great weight is still attached in the public debate to the geographical periphery.

Discussing peripheral phenomena in a large urban agglomeration like, for instance, New York, might seem presumptuous. However, if we do not merely attach labels expressing geographical distance to peripherality and centrality but also give them meanings which express social and emotional belonging, parts of large urban agglomerations may appear significantly more peripheral than many rural areas.

It should be emphasised that there is a great danger in generalising urban agglomerations. Thus, we need greater understanding of the whole range of parts that go to make up an urban structure and of what they mean to the people who live and work there. A transition to a post-industrial society means that the value structure is continually changing, with a silent revolution resulting in material values of security being replaced by immaterial values of liberty and freedom (Fürth, 1994). The immaterial factors determining the welfare and quality of life of the individual thus gain increased significance and must be taken into account to a greater extent.

Given the rapid urban expansion that is taking place throughout the world, it is estimated that five urban agglomerations will have increased their population by more than 50% in the decade from 1990 to 2000. Population-wise the largest urban agglomerations in 2000 will be Mexico City, Sao Paulo and Tokyo. In total the 28 largest urban agglomerations in the world will, it is estimated, have increased their population by an average of 31% from 1990 to 2000 and will together have a population of almost 370 million. In a European perspective, however, the rate of urbanisation appears relatively moderate. It has almost ceased in the central areas of the continent and it is estimated that in 2000 Moscow will be Europe's largest city, with a population of about nine million (World Urbanization Prospect, 1991).

I believe therefore that it is of some considerable importance that, in a society where we are rapidly reaching a 50% level of urbanisation in the world, we should increasingly discuss the urban microperipherality, which is becoming more and more apparent. However, it is not a question turning our backs on today's rural areas and the problem of peripherality which exists there and which will probably grow as urbanisation is still in a state of rapid expansion in many parts of the world. It is a matter of being able draw lessons regarding the knowledge and competence that exists in today's peripheral rural areas and apply them to tackling the new form of centre-periphery problem. In studies of urban transformation it is essential to bring together global and local issues since the factors influencing development in individual urban areas are often a reflection of the economic activity in the national and international arena. This is not least true of many of the rapidly expanding third-world economies, one result of which is an explosive rate of urbanisation.

Summary

Are we facing the fact that a new form of Mega-city periphery is emerging which differs from the periphery we have and experience in many rural areas?

Periphery is a concept that is frequently associated with rural areas and the problems that exist there in the form of high unemployment, low formal education, high level of sick leave, low average income, poor public transport. Rural postmen, shops only open between 10 am and 4 p.m., closed schools and problems at the factory. It is on this basis that rural areas and peripheral regions are branded as so-called problem areas. Is it possible and relevant then to speak of the periphery and peripheral phenomena in population centres and urban agglomerations? In the present study it is argued that it is important to introduce a further dimension into the centre-periphery debate, focusing on the difference between material and immaterial peripherality and the difference between material and immaterial centrality. Material peripherality refers primarily to a physical sparseness of buildings and service functions. This type of peripherality is measurable and can thus be analysed with the aid of quantitative methods. Immaterial peripherality is rather what might be termed mental or emotional peripherality.

The empirical material in the study is drawn from New York City (NYC). My purpose in studying NYC and, in particular, Manhattan is that this enables me to discuss the differences and similarities between adjoining urban districts from a centre-periphery perspective.

In the study of the twelve districts that make up Manhattan, a distinct picture emerges of a segregated urban centre where the differences in actual living conditions are considerable. The statistics on which the study is based only reveal a material side of what may be classified as central and peripheral respectively. However, given the variables used in the study, it is possible to form a picture of which areas would be classed by many people as 'good', 'less good' and 'bad'. Many of our evaluations of a region or perhaps a specific district are based on the stereotypical 'images' we form with the aid of statistics, the media and rumour.

We seldom have 'inside information' from those living in the areas in question and we often base our notions on secondary sources. It is essential to stress the contrast between the way an 'insider' perceives a place and the conventional manner in which an 'outsider' describes it.

Further, it is important to emphasise the danger of generalising urban agglomerations. We must gain a greater understanding of the whole range of components in an urban structure and what they mean for the people who live and work there. The transition to a post-industrial society entails continual changes in value structures, with a silent revolution resulting in the material values of security being replaced by immaterial values of liberty and freedom. The immaterial factors determining the welfare and quality of life of the individual thus gain added importance and must be taken into consideration to a much greater extent.

References

Blom, T. (1996a), *Perspektiv på kunskap och utveckling. Om attityder till högskoleutbildning i några perifera regioner,* Meddelanden från Göteborgs universitets geografiska institutioner, Series B 89, Göteborg, (Also published by Högskolan i Karlstad in the series Forskningsrapport Samhällsvetenskap, no 96:3, Karlstad).

Blom, T. (1996b), *Periferi och centralitet. En fråga om perspektiv?* Arbetsrapport 96:11, Samhällsvetenskap, Gruppen för regionalvetenskaplig forskning, Högskolan i Karlstad, Karlstad.

Blom, T. (1997), *Mikroperiferitet i storstadsregioner – ett ökande utvecklingsproblem.* Carlstad University Press, Vålberg, (Also published as Arbetsrapport 97:4, Avd. för Geografi och Turism, Högskolan i Karlstad).

Buttimer, A. (1981), 'On People, Paradigms and 'Progress' in Geography', in D.R. Stoddart, (ed.), *Geography, Ideology and Social Concern,* Oxford.

Cadwallader, M. (1996), *Urban Geography, An Analytical Approach,* Prentice Hall, Upper Saddle River, NJ.

Cater, J. and Jones, T. (1989), *Social Geography. An Introduction to Contemporary Issues,* London, Arnold.

Community District Needs, Manhattan (1996), *Fiscal Year 1996/The City of New York, Office of Management and Budget, Department of City Planning.* NYC DCP #94-22, New York.

Fürth, T. (1994), I FRAMTIDER nr1/94, Institutet för framtidsstudier, Stockholm.

Giddens, A. (1994), *Sociologi*, Studentlitteratur, Lund.

Hayter, R. and Barnes, T.J. (1992), 'Labour Market Segmentation, Flexibility and Recession: A British Colombia Case Study', *Environment and Planning C: Government and Policy*, Vol 10.

Knox, P.L. (1994), *Urbanisation*, Prentice Hall, Englewood Cliffs, NJ.

Law, R.M. and Woch, J.R. (1993), 'Social Reproduction in the City: Restructuring in Time and Space', in P.L. Knox, (ed.), *The Restless Urban Landscape*, Prentice Hall, Englewood Cliffs, NJ.

McDowell, L. (1995), 'Understanding Diversity: The Problem of of/for ''Theory'''', in R.J. Johnson, P.T. Taylor, and M.J. Watts, (eds.), *Geographies of Global Change*, Blackwell, Cambridge, MA.

Mehretu, A., Pigozzi, B.W. and Sommers L.M. (1996), 'Issues of Urban Marginality in the Greater Detroit Area', G. Jones and A. Morris (eds.), *Issues of Environmental, Economics and Social Stability in the Development of Marginal Regions: Practices and Evaluation,*. Dept. of Geography University of Strathclyde and Dept. of Geography University of Glasgow, Scotland.

SOU 1997:118, *Storstadskommitténs betänkade Delade städer,* Fritzes, Stockholm.

Wille, W.W. (1993), *New York – Resa på annat sätt*, Alfabeta Bokförlag, AB Fälths tryckeri, Värnamo.

World Urbanization Prospect 1990 (1991), *Estimate and Projections of Urban and Rural Populations and of Urban Agglomerations*, United Nations, New York.

16 Space, behaviour and marginality

PAULO NOSSA

Introduction

This article does underline the clinical or etiological aspects of HIV infection (Human Immunodeficiency Virus). Those have been largely debated in 80's and early 90's, and set up a medical cluster in the global programme against HIV/AIDS. Therefore it looks more appropriate, and more geographic, to discuss the spatial spread of HIV, the multiple territory factors which could be involved on HIV control or diffusion.

Conceptual changes

Later in 80's, some ambiguous convictions had been removed, some of them point out sexual preferences as a single motive to the infection, swayed by prejudice, emerging rough designations such as: 'cancer gay' and 'pink fever'.

Systematically the aetiology became clear with the spatial diffusion of AIDS (Acquired Immunodeficiency Syndrome), touching different social classes by different ways. Firstly, homosexuals with multiple partners and secondly, sex workers and migrants, specially young Haitians in the USA, which were involved in commercial sex through tourism activity in there home lands, both groups were a easy target, social excluded or stigmatised by poverty or illiteracy (Gould, 1993).

In 1989, medical experts and WHO (World Health Organization), struggle this prejudice, which added more and more difficulties to this uncovered population. In official policy, some designations had been changed: risk group designations were changed by behaviour risk group, identifying some aspects of individual behaviour as a determinant on HIV infection. After, in 1992, some social researchers in developing countries as Mozambique, introduced a straight concept – 'high risk situation': as all the situations which allow social, economics or political changes, able to increase the risk of becoming infected with HIV, underlining the regional or local context (Bartlett, 1998).

By this way, they hope to produce more accurate answers and preventive policies according to the needs of exposed populations, specially named 'hidden populations', as: individual or groups which are not able to claim their social or human rights, according the status or feeling of exclusion.

In this context we can summarise the main factors related with high-risk situation: war and guerrilla war, migratory chain, economic breakdown, and global tourist demand:

1. War and guerrilla war.
   - crush of social and health care equipment;
   - aleatory displacement of population;
   - overloading areas of refuges;
   - human rights violations, special sexual abuse;
   - increment of wilfulness in sexual negotiation;
2. Social disruption.
3. Migratory chain.
   - social exclusion;
   - social and health care precariousness;
   - increase of demand/supply on commercial sex;
4. Increasing possibility of spread infection diseases in main rotes of migration (including STD [Sexually Transmitted Diseases]).
5. Economic breakdown.
   - cuts in social and health care programmes affecting 'uncover population' and primary prevention;
   - increase of migrant chain;
   - depart from rural areas to poor urban periphery;
   - increase of homeless population;
   - low priority to health care;
   - increase of demand/supply of commercial sex, low capacity to negotiate safer sex;
6. Low self-esteem.
7. Global tourist demand.
   - diversified tourist sector with increasing demand/supply of exotic places including commercial sex;
   - high percentage of seasonally migrant population and increase of spread of STD, including HIV;
8. Exploration of low incomes between local population, which can supply commercial sex with light social, penalises.

These and other factors had received lower consideration on drawing educational and sanitary programmes in developing countries, establishing one of the most clamour inequality on the budget prevention policies between poor and riches countries. Separately or in mass, these pictures had been operated as hide co-factors in the diffusion of pandemic, result of political and economic adjourn which correcting measures were claimed for so long time. According data published by UNAIDS (United Nations Programme on HIV/AIDS) we can find to 1997, newly 5.8 million cases of HIV infection, 89% of global cases are placing in sub-Saharan Africa and developing countries of Asia, increased by 8.2 million of AIDS orphans,[1] all over the world, since the beginning of the epidemic (Figure 16.1).

After Vancouver World Conference on AIDS, 1996, scientific community were allowed to care the HIV infection no more as a lethal stage, changing to a chronically disease, using the new and expensive antiretroviral therapy (substances used against retroviruses), which increase the surviving with more quality of life. The access to this new and yearned treatment clearly show the gap between developed and developing countries, above marginalized groups, and peoples of poor nations, several times we can consider a whole continent. This set up a brutal and sever way which offence the basic human rights and added more inequality: choosing between a early death to a marginalized population in developing countries, or keep the disease in a chronically stage, which demand continued and expensive health care in developed countries (UNAIDS/WHO, 1998). The last World Conference on AIDS, Geneva, July 1998, show this inequality or weak situation and imposes a challenge: Bridging the Gap (UNAIDS/WHO, 1998). Bridging the Gap in side the countries adopting more capable educational and prevention programmes, which consider the particular needs of 'hidden populations'.

Bridging the Gap between the reach north and poor south, developing an international therapeutic solidarity fund, across international mobilisation focusing on the main objective of access to expensive treatment for people living with HIV/AIDS in developing countries and drawing a fit policies in prevention area, enfolding social aspects. This International Therapeutic Solidarity Fund, proposed by the President of Republic and the Secretary of State for Health for France, during the 10th International Conference on AIDS in Africa, should had been subscribe by European Union in December 1997, according UNAIDS programme, and had been also examined by G8 in Birmingham meeting May 1998. In this point the subscriptions leaders sharing four important aims:

1) = North America, 860,000; 2) = Caribbean, 310,000; 3) = Latin America, 1.3 millions; 4) = Western Europe, 480,000; 5) = North Africa & Middle East, 210,000; 6) = Sub-Saharan Africa; 21 millions; 7) = Eastern Europe & Central Asia, 190,000; 8) = East Asia & Pacific, 420,000; 9 = South & SE Asia, 5.8 millions; 10) = Australia & New Zealand, 12,000.

Figure 16.1   Adults and children living with HIV/AIDS, end of 1997

Source: UNAIDS/WHO 1998.

1. An ethical consideration, which rules out indifference facing the suffering and death of million of people, and which refuses the logic of a two speed epidemic, sacrificing the people living with HIV/AIDS in the countries in the south.
2. Public health considerations based on the fact that there cannot be sustained control of the epidemic without real complementarity between the prevention policy, and provision of medical care.
3. The necessity of preserving the progress made in terms of development. The epidemic reduces the hard-won gains in some countries, bringing them back down to the socio-economic level of the 1960's. The micro, then macro-economic demographic impacts are leading to deterioration in national economies. Combined with major social and cultural consequences, they accentuate inequality, increasing the risk of exclusion (especially for women) and inevitably compromise the political stability of countries.
4. Concerns over geopolitical security, because in the eras of globalization, this huge epidemic linked to communication, exchanges and population movements, has already affected international balances and will do so even more in the future.

In human rights field, fighting this deep inequality is an imposed duty, which should take, as soon as possible, an international dimension. According UNAIDS Report, June 1998, for developing countries, HIV diffusion has no simple explanations:

'Globally, it is certainly the poorer and less educated who are feeling the brunt of HIV epidemic. Nevertheless, the epidemic has spread in different ways and through different groups of people in different parts of the world. Neighbouring countries often have very different epidemics. In addition, even within a single nation, HIV can strike different populations or different geographic areas in dissimilar ways, ways that may change over the course of time.'

Recording what we have already appointed, special related with the 'high risk situations', we should demonstrate in a clear way that educational programmes are the starting lever in AIDS control around the world:

'Better-educated people have better access to information about HIV, how it is transmitted, and how it can be avoided. On top of that, better educated people are more likely to have better paid jobs, and can afford the sorts of goods and services that allow them to act on their AIDS knowledge' (UNAIDS/WHO, 1998).

As we can see, just a collective effort to increment a global consciousness can help us to delete this big gap of inequality, reducing this nearly medieval 'triumph of death'.

Note

1. Defined as children who lost their mother or both parents to AIDS when they were under the age of 15 (WHO).

References

Bartlett, John (1998), *1998 Medical Management of HIV infection*, John Hopkins University, Department of Infectious Diseases.
Gould, Peter (1993), *The slow plague: a geography of the AIDS pandemic*, 1st Edition, Blackwell Publishers.
Grmek, Mirko (1990), *Histoire du SIDA*, 2 Édition, Édition Payot, Paris.
Piot, Peter, (1998), The science of AIDS: a tale of two Worlds, *Science*, Vol. 280, pp.1844-1845.
UNAIDS/WHO (1998), *Report on the global HIV/AIDS*, June 1998, Geneva.
Zwi, Anthony & Cabral (1992), Identificar 'situações de alto risco' para prevenir a SIDA, *British Medical Journal*, Portuguese Edition, Vol. I, nº 3, pp.232-236.

Part 3 – Development in margins and peripheries

17 A pink invasion into the Dutch periphery

JAN H.M. MAAS AND JOHAN WISSERHOF

Introduction

This paper investigates a consequence of globalization on marginal regions in the Netherlands. The ongoing expansion of the pig sector in Dutch agriculture is representative of the process of globalization in this case. This expansion is now spreading into peripheral rural areas that had been saved from (the various negative side effects of) intensive pig raising until recently. The interesting point in the 'pink (pig) invasion' is that the regions studied, despite their marginal position, seem to be able to resist the invasion by the pig sector, as we will show below. A process of regionalisation occurs in the face of globalization.

Due to its controversial nature, the pink invasion allows us to gain insights into driving forces of regional dynamics, which would otherwise remain latent. The controversy between the supporters and the opponents of the pink invasion clearly highlights the various factors and actors involved in the development of the regions studied, the ideas, interests and institutions, the economic and political power relations, etc.

Our approach is to study regional dynamics from the perspective of geographical scale. This is to say that we focus attention on aspects of scale of the pink invasion, particularly the 're-scaling' of the social activities involved. We believe this is a fruitful approach to understanding regional dynamics in the context of globalization, which is a process of re-scaling itself. We hypothesise that regional dynamics or the *restructuring* of rural regions today can be understood as the *re-scaling* of the social activities with regard to those regions. Rescaling may occur both in an 'upward' direction (e.g., the internationalisation of intensive animal husbandry) and in a 'downward' direction (e.g., the regionalisation of arable farming).

Study regions and outline of the paper

The study regions are Zeeland Flanders and Eastern Groningen (Figure 17.1). These regions are peripheral border areas in the Netherlands. They both have a surface area of about 700 km², a population of approximately 100,000 inhabitants, and arable farming as the main form of land use (mainly grains, potatoes and sugar beets).

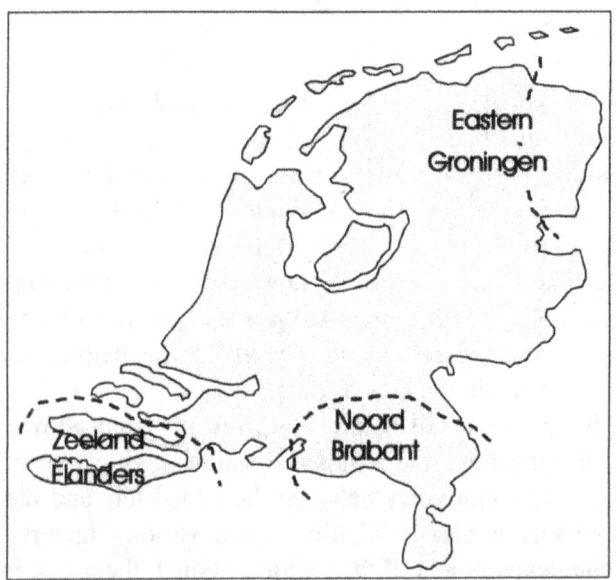

Figure 17.1   The study regions in the Netherlands

The pink invasion into these regions can be explained mainly by means of the developments within two sectors of Dutch agriculture. In the regions of origin, the problems caused by the expansion of animal husbandry and its high concentration in relatively small regions in the South, Centre and East of the Netherlands act as *push-factors*. In the immigration areas in the Southwest and Northeast of the Netherlands, the stagnation of arable farming as a consequence of the low profitability of many crops functions as the main *pull-factor*, all the more so because in the Dutch agricultural context arable farming can be qualified as an extensive sector.

In the following sections, first the push- and pull-factors are described in terms of the internationalisation of the Dutch pig sector and the stagnation of arable farming, respectively. Then, the resulting pink invasion into the arable farming regions is analysed in greater detail. The last section summarises the main conclusions drawn from the analysis.

The internationalisation of the Dutch pig sector

The primary cause of the 'pink invasion' is the growing internationalisation of the Dutch pig sector. Until the second half of the 19th century, the pig sector mainly had local and regional relationships, and hence its development depended on endogenous forces. Pig raising was found in the Low Netherlands (especially the West), as the big cities here offered a market for meat and because pigs were able to transform the residues of the regionally important dairy industry into valuable products. In the High Netherlands (the South and East), pigs were not only kept on the numerous small farms to process otherwise useless plant residues, but on these poor sandy soils they were also indispensable for the production of manure. It was here that the pig concentration areas came into being at a later stage. Conversion to pig farming proved to be one of the most suitable and easiest means of intensification for the numerous small farmers in these regions with traditional gavelkind inheritance of land.

Internationalisation started at the end of the 19th century, when the Netherlands opted for free trade, in contrast with the surrounding countries, which gave priority to protection of their farmers against the consequences of the agricultural crisis of the 1880s. Cheap grains for animal feed could be imported into the Netherlands from production countries overseas and animal products could be sold in the neighbouring countries. In this way, some hundred years ago the foundation was laid for the functioning of the Dutch pig sector as an international 'upgrading' industry, importing raw materials from other continents, processing them by means of the livestock and exporting the products to other European countries. This principle has remained the same until now, though in the meantime great changes have taken place in the dimensions of the activities, the (geographical) structure of the sector in the Netherlands, the choice of the countries providing raw materials for feedstuffs and the extension of the foreign market, as we will see below.

In the first half of the 20th century, the pig population of the Netherlands grew gradually, and with some ups and downs, from 750,000 in 1900 to 1.9 million in 1950. The strong increase in numbers began around 1965. There were 3 million pigs in 1960, 5.5 million in 1970, 10.1 million in 1980 and nearly 14 million in 1990 (see Figure 17.2). The 1997 epidemic of Classic Swine Fever only led to a slight decrease to 13.3 million in 1998, even though some 10 million pigs and piglets were killed.

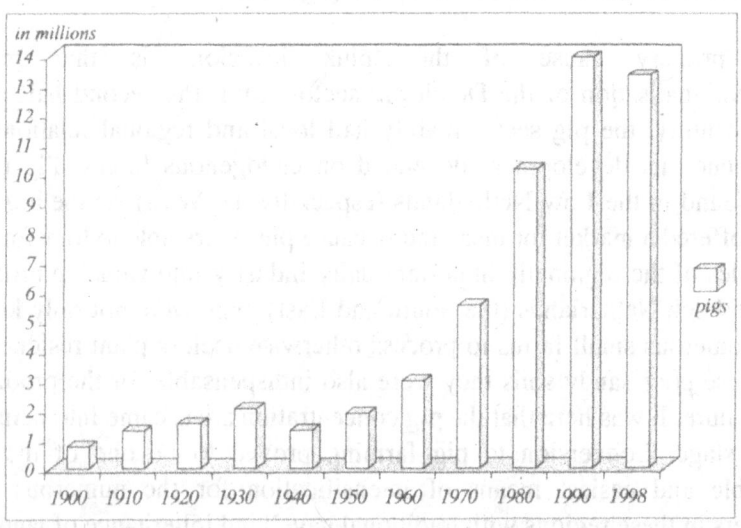

Figure 17.2 Millions of pigs in the Netherlands, 1900-1998

Source: CBS, Agricultural censuses.

With the growth of the pig sector and its concentration in relatively small areas, localised or geographical agro-complexes arose where all activities of the pig production chain were located close to one another: pig farms (which became ever larger and more specialised), input providers (compound feed producers, stable constructors), buyers of products (slaughters, livestock traders) and service people (veterinarians, bankers, transporters). More than half of the pigs in the Netherlands can be found in Noord Brabant now and one third in the concentration areas in the Centre and East.

The complexes could expand so quickly thanks to the favourable circumstances on the input (providing of raw materials) and output (marketing of products) sides of the production chain and due to the agglomeration and urbanisation economies brought about by the geographical concentration of the various parts of the production chain (Maas, 1994; Maas and Segrelles, 1996; 1997). The geographical situation and the seaport infrastructure allowed shipping of raw materials for animal feed in great quantities and at low prices from all over the world to the Netherlands. Initially, mainly grains from Russia and North America (maize) were imported. Later, countries in South America and Southeast Asia also provided raw materials, which increasingly came to consist of non-grain ingredients. As a consequence of a gap in Community legislation, non-grain ingredients for animal feed produced in third countries could be brought into the Community practically free of charge. Now, nearly 89% of the raw materials for the Dutch compound feed industry are imported, of which only one third is derived from other EU countries. The sizeable amounts of raw materials imported through the port of Rotterdam offered the Dutch pig sector a significant cost advantage in comparison with other production areas of the EU, although this advantage has diminished by the lower grain prices following the reform of the Common Agricultural Policy in 1992. Currently, pig feed in the Netherlands is made of 30% tapioca, Thailand acting as the main producer. Grains (European countries), rape-seed shred (Germany) and shred of sunflower seeds (Argentina) each contribute for about 15%. France contributes peas and several American countries soybean shred (*Boerderij*, 3-3-1998).

In addition to the opportunity of importing cheap raw materials, favourable infrastructure, geographical, sectoral and societal circumstances (education, transportation network, regional concentration, co-operatives, and public support) contributed to the success of the Dutch pig production chain. Some of these were also very beneficial for the marketing of products abroad. In the first half of the 20th century, the export products of the Dutch pig sector mainly found a market in the nearby industrial regions of Germany and England. This situation changed after the start of the European integration with the Treaty of Rome in 1957. The ongoing expansion of the European Community (later the European Union) implied an ever greater potential market for products from the pig sector. Before the recent outbreak of swine fever, the Netherlands were the greatest exporter of live animals and, after Denmark, the second largest exporter of pork, and 67% of the total production value of the Dutch pig sector was exported (in 1995). Only small quantities went to countries outside the EU such as Japan, the United States and Eastern Europe. Ninety-nine percent (99%) of the pork export went to EU countries, of which Germany bought 35%, Italy 24% and the United Kingdom 15%. Live animals (piglets and pigs for slaughter) mainly went to the neighbouring countries Germany (48%) and Belgium (13%), but not less than 18% and 11% was bought by Italy and Spain, respectively, and a small part went to non-European countries such as Japan (Van Gaasbeek, et al., 1993; LEI-CBS, 1997).

A conclusion that can be drawn from the foregoing is that there is a great difference in geographical scale between the input and the output sides of the Dutch pig sector and that input and output did not have the same development of geographical scale in the course of time. Raw materials for animal feed have already been obtained from outside Europe for a hundred years, but more recently they are coming from more countries, which are, located world-wide. The products of the pig sector are sold nearly entirely inside Europe, and countries at a greater distance from the Netherlands are accounting for an ever-greater share.

A second conclusion is that the regional concentration in geographical agro-complexes carries with it both advantages and disadvantages. Agglomeration and urbanisation economies, largely due to the short distances between the various parts of the production chain, make for lower production costs, which provide a favourable starting-point for competition on foreign markets. On the other hand, the higher risk of the spread of contagious animal diseases and, above all, environmental pollution is clear examples of agglomeration disadvantages. Since an average of 60% of the manure cannot be used on the farm where it is produced, the huge expansion of intensive animal husbandry causes an ever-greater manure surplus. Part of it is trans-ported by road to the arable farming regions in the north and the south-west of the country. Environmental problems are aggravated even more by the great many dairy farms and other branches of factory farming (poultry, fat-tening calves) in the concentration areas. The negative consequences of the manure surplus, such as phosphate accumulations in the soil, ammonia emis-sions in the air and smell pollution, increasingly worried public opinion and politicians. This led to the first legal provision to impose restraints on intensive animal husbandry in the concentration areas in 1984. Later, at the level of the EU regulation arrangements were also made to protect groundwater and surface water from nitrate pollution. Therefore, this problem of Dutch animal husbandry has a European dimension as well. Another agglomeration diseconomy of the concentration areas is the high price of land. Therefore, the purchase of land to improve one's 'manure rights' (surface-determined manure production quota) is expensive. Between June 1997 and June 1998, the average price for arable land in the concentration areas of the South, Centre and East was about 60,000 guilders per hectare; in Zeeland Flanders, this was about 40,000 guilders and in Eastern Groningen 35,000 (*Boerderij,* 30-6-1998). Such regional differences now also appear for the transferable 'pig rights' (quota for pig keeping) which were established in the Dutch national Restructuring Law of 1998 to replace the manure rights (*Agrarisch Dagblad*, 18-7-1998).

Despite increasing problems, the pig sector has continued to expand in the past decades. Public opinion, which used to be quite positive about the sector due to its high contributions to the national economy and the balance of trade, changed rapidly as of the beginning of the nineties. The outbreak of the swine fever epidemic in February 1997, which ravaged the sector for more than a year, marked a turning point. The Minister of Agriculture announced far-reaching measures. One of these is the above-mentioned Restructuring Law, aiming at reductions of the 'pig rights' up to 25% in the year 2000 if the farm does not meet certain ecologically sound criteria. Next, there is the Pig Order (*Varkensbesluit*), meant to improve sanitary and living conditions. A Reconstruction Law is being prepared, which will split up the concentration areas into separate clusters to reduce the risk of the spread of contagious diseases.

The reactions of the pig farmers to the increasing problems facing the sector (growing agglomeration disadvantages, deteriorating image, social resistance, and restrictive public measures) vary greatly. A considerable minority tries to prevent changes and to stop public measures by obstruction. A large part adapts to the new reality by investing in ecologically sound means of production (like stables that reduce ammonia emissions) or ways of production (free-range pigs). Others seek a solution by leaving the concentration area. Often, these farmers have large farms, and sometimes they only move part of their production capacity to another location, in particular the production of pigs for slaughter. The new location can be situated abroad or at home. After the *Wende* of 1989, a number of Dutch pig farmers went to Eastern Germany; currently, countries such as Canada, the United States, Poland and Spain arouse more interest. Inside the Netherlands, pig farmers are looking for areas with little livestock and without a manure surplus, that is, regions with arable farming. Moreover, they prefer thinly populated regions with a peripheral location in the country. Here, less land is being claimed for non-agrarian purposes, and the probability of protests against nuisance is smaller. The lesser demand for land and the low profitability of arable farming here mean that prices of farms and land are relatively low. As transport of manure is not necessary, production costs are also lower for the pig farmers.

In sum, since the end of the 19th century, the Dutch pig sectors has enlarged its action space to the entire world, first by the acquisition of feedstuffs, then by the marketing of its products, and now even by the location of production units.

Stagnation of arable farming regions: opportunities for pig farming

The situation in the pig sector in the concentration areas as described above is the direct cause of the 'pink invasion' into certain regions in the North and Southwest of the Netherlands. This invasion is made possible, however, by the weak position of agriculture, particularly arable farming, in the struggle for space in these regions. Internationalisation and globalization have also influenced Dutch arable farming, but contrary to the pig sector this process, in the end, has led to stagnation and decline. It started with the Agricultural Crisis of the 1880s, when the importation of cheap grains from overseas caused a considerable drop of wheat prices. The arable farmers responded with modernisation of their farms on the one hand, by introducing new forms of organisation (co-operatives), methods of production (mechanisa-tion) and means of production (artificial fertiliser), and, on the other hand, by introducing new crops such as sugar beet, starch potatoes and flax, which could be sold to the emerging regional agro-industry. Until the Second World War, local and regional actors, i.e., the home market and the policy of the national government mainly determined the development of arable farming.

After the Second World War, the modernisation of arable farming accelerated considerably. As a consequence, farming became more and more dependent on machines, fuels, artificial fertilisers and chemicals that were produced in other parts of the country or abroad. The number of agricultural workers declined sharply. After the founding of the European Community, agricultural policy moved from the national to the international level. The crop plan of Dutch arable farming became less varied as it specialised mainly in crops supported by the market and price policy of the Communal Agricultural Policy (CAP): grains, sugar beets and starch potatoes. Thanks to the high support prices, arable farming generally prospered in the 1960s and 1970s. This situation deteriorated when the towering production surpluses and soaring budget expenditures forced things to change. Financial support was limited (lower support prices, stabilisers) and production restricting measures (quotas, set-aside) were introduced. When these measures turned out to be insufficient, more radical reforms were implemented in 1992 (the MacSharry Plan), with important measures for arable farmers such as a shift from price support to income support, lowering of the intervention price for wheat by 30% and sharper conditions for the set-aside of arable land.

In the 1980s, arable farming began to stagnate and decline as many farmers gave up their businesses. Profitability decreased as a result of lowering receipts and rising expenses. The lower receipts were not only brought about by the decreasing support of the CAP, but also by the liberalisation of international trade (GATT agreements), leading to stronger competition from producers inside the EU and in third countries. Grain prices, which had been increasing since 1950, began to decrease in 1983, bringing the price nearly back to its post-war level in 1998 (*Agrarisch Dagblad*, 10-7-98). The higher expenses are caused, among other things, by environmental protection measures of the national government and by rising prices of the means of production.

In the 1990s, the growing international competition makes clear that Dutch arable farmers have a weak position on the EU market for most of their products and that their position on the world market is even worse. Their farms are relatively small and they are specialised in crops with a market surplus or for which such a surplus can be expected before long. On the other hand, they cannot make an appeal to the special EU support funds as their farms are modern and natural production conditions are not unfavour-able. The deterioration of the financial results of the Dutch arable farming sector is clearly visible from the annual average income per 100 guilders of production costs. Until the end of the 1970s, this indicator was (far) higher than 100, but since then only a few years show such a figure. For the period 1991-1996, the average was only 89. Regarding the struggle for space between arable farming and the pig sector, it is interesting to see that, according to the same indicator, the results of arable farming between 1994 and 1998 decreased with 17 guilders to 87 while those of the pig sector rose by about the same sum to 102 (LEI-DLO, various years).

To protect arable farmers and arable farming regions from impoverishment, there is a constant search for new or alternative economic activities on farms and in the region. At the farm level, new activities are introduced to exploit the available labour force and the fixed means of production. Farmers are introducing new activities such as other crops (hemp, chicory), a second or third agricultural branch (vegetable growing in the open, poultry or pig farming) or another way of production (organic farming). Others seek to supplement their income by non-agricultural activities such as 'gate' sale of agricultural products to consumers, offering accommodation (campsite, rooms) and recreation facilities (horse riding, canoeing), or working outside the agricultural sector. Especially in Zeeland Flanders, many farmers are offering tourist facilities. At the local and regional level, authorities are encouraging the settlement of farm types that offer better perspectives than arable farming, like horticulture under glass and dairying. Generally, settlement does not cause problems or resistance on the part of the population of the regions concerned, with one exception: the settlement in Zeeland and Groningen by pig farms from the concentration areas. Especially pig farmers from the province of Noord-Brabant, the major concentration area, are making use of the weak position of arable farming in these provinces, meaning that agricultural land use is rather extensive, that profitability is insufficient and that, in contrast to the greater part of the Netherlands, these regions do not have a manure surplus. This situation is very attractive to pig-farmers from the concentration areas. The resulting 'pink invasion' into Zeeland and Groningen will be analysed below.

We conclude that the stagnation and decline of arable farming and arable farming regions is mainly caused by the growing competition resulting from *internationalisation* and *globalization* of agricultural markets and agricultural policy. As with the pig sector, this process started more than a century ago and accelerated after the 1950s as part of the European integration. The first decades of the CAP turned out to be favourable to Dutch arable farming, but by the end of the 1970s cutbacks in EU support and the growing liberalisation of the world market began to weaken its competitiveness. Arable farming and the agricultural regions specialised in this sector began to decline. To turn the tide, farmers as well as regional authorities are searching for alternative activities based on local and regional resources. This 'endogenous development' (Van der Ploeg and Long, 1994) implies a partial *regionalisation* of arable farming. In this way, the regions are trying to stop the negative developments caused by the processes of internationalisation and globalization. The resistance in these regions to the pink invasion also fits in with the movement for endogenous development, as the inhabitants fear that large-scale pig farming will frustrate or complicate the realisation of promising regional initiatives such as (agro-)tourism and recreation, (agricultural) nature conservation and organic farming.

The pink invasion

The impossibility for the pig sector to expand further in the concentration areas (push-factor) together with the existence of regions showing a stagnation of arable farming and a minor development of animal husbandry (pull-factor), as described in the preceding sections, bring about a geographical shift of the pig production capacity. Although this 'pink (pig) invasion' is a movement at the inter-regional level, its origins lie in the specific situation in certain Dutch regions on the one hand and in developments of international and global political and economic relationships and in world trade on the other, as has been explained before. The pink invasion thus illustrates that the restructuring of the rural regions of origin and destination is linked to national and international re-scaling of agriculture.

Migrations of pig farmers are taking place from all concentration areas in the South, Centre and East of the Netherlands on the one hand to all arable farming regions in the Northeast and Southwest on the other, but here we focus on the main movements, the 'invasions' of Zeeland Flanders and Eastern Groningen (the most peripheral regions of the Netherlands) by pig farmers who originate mainly from the eastern part of the province of Noord-Brabant (the major concentration area; see Figure 17.1).

As primary data on the migration of pig farms in the Netherlands are lacking, the 'pink invasion' can only be traced indirectly. First, there are the results of the Yearly Agricultural census (Landbouwtelling) of the Central Bureau for Statistics (CBS). On the basis of indexes calculated from the number of pigs, the developments in Zeeland Flanders and Eastern Groningen can be compared with those in the Netherlands in general. If we take the numbers in 1980 as the base (= 100), the stock of pigs in the Netherlands had risen to 150 in 1997, in Eastern Groningen to 184, but in Zeeland Flanders it decreased to 87 (see Table 17.1). So it appears that there has not been an invasion of pigs in Zeeland Flanders, only in Eastern Groningen. But neither for Zeeland Flanders nor for Eastern Groningen can this conclusion be drawn unconditionally as it is unknown to what extent the pig stock that was already present in these regions changed between 1980 and 1997. Even so, the absolute numbers of pigs in the two regions in 1997 are still insignificant compared to the rest of the country: 86,308 in Eastern Groningen and 34,889 in Zeeland Flanders, together less than 1% of the national stock of more than 15 million pigs and less than 2% of the pigs in the province of Noord-Brabant.

Table 17.1   Development of pig farming 1980-1997, (1980=100)

	Netherlands	Zeeland Flanders	Eastern Groningen
Stock of pigs	150	87	184
Pigs for slaughter	142	172	223
Breeding pigs	126	47	108
Stable capacity	142	199	201

Source: CBS Agricultural censuses.

The change in the composition of the stock of pigs offers a second indicator for a geographical shift, as it is known that big pig farmers move the fattening of pigs for slaughter to the branch farm(s) they start in another region. Fattening requires less labour, knowledge and investments than the breeding of new crossings or the reproduction of pigs. Therefore, the farmers keep these more gainful activities on their original farm(s) in the concentration area. So relocation not only leads to an increase in scale of the farm itself. It is also a process of spatial scaling up as selected activities are moved out, just like manufacturing industries can relocate particular production processes to places which have a special location advantage. Now the CBS census data show that the stock of pigs for slaughter in the Netherlands increased from 100 in 1980 to 142 in 1997, in

Eastern Groningen to 223 and in Zeeland Flanders to 172. For breeding pigs, the indexes are 126 for the Netherlands, 108 for Eastern Groningen and 47 for Zeeland Flanders. This means that, between 1980 and 1997, both study regions were strongly specialising in pigs for slaughter.

Information from municipalities and the Regional Inspection of Environmental Hygiene (RIMH) on applications and grants for building permits, besides data on the development of stable capacity, offer a third indication for the pink invasion. According to data from municipalities and the RIMH, applications and grants for permits in Zeeland Flanders and Eastern Groningen since 1990 concerned stable capacity for hundreds of thousands of pigs, while completion of stables concerned a total of several tens of thousands of places. It seems probable that many applications were made by farmers applying in different municipalities in order to raise their chances of getting the desired capacity, as the numbers applied for are proportional neither to the numbers of realised capacity nor to the growth of stable capacity, as is evident from the CBS agricultural censuses. According to this source, capacity in the Netherlands rose from 6.4 million pig places in 1980 to 9.1 million in 1997, an increase of 42%. In the two study regions, the capacity during this period doubled exactly, to 53,431 places in Eastern Groningen and to 29,355 in Zeeland Flanders. Only 14,501 and 4,354 of these numbers, respectively, were realised after 1990.

The absolute and relative figures presented above do not indicate a massive migration of pig-farming activities nor do they show an invasion. In Zeeland Flanders, there has certainly been no real invasion until now, and in Eastern Groningen there might have been an invasion of modest proportions. Moreover, a substantial migration in the future seems improbable now, as the Reconstruction Law for the pig sector that is being prepared intends 'to keep clean areas clean'. Expansion can only take place on the basis of permits that have already been allocated, but it is unknown what capacity is still in the pipeline. Therefore, it seems likely that the strong public resistance in Zeeland Flanders and Eastern Groningen against a pig invasion is inspired more by the great number of applications and uncertainty regarding the numbers that might finally be realised than by any factual expansion of the pig stock. The fact that some 'pig barons' from Noord-Brabant use all (legal) means to realise their plans also strengthens the opposition.

Although the factual dimensions of the 'pink invasion' are not well known and are most probably overestimated, it has become an important social and political issue in the regions concerned during the past decade. In the 1980s and 1990s, Zeeland Flanders and Eastern Groningen became the arena for great many actors involved in this process. They can be divided into opponents of the pink invasion, mainly living in the regions concerned, and supporters, mainly from outside. In addition, there are actors like public authorities and farmers' organisations who, as a consequence of their position and interests, do not side unambiguously with one or the other position (Olijve, 1998).

Obviously, the instigators of the pink invasion, the pig farmers searching for alternative locations for their activities, belong to its *supporters*. There are different groups such as starters and pig farmers who have to give way to expanding cities or nature conservation areas. But the most eye-catching group consists of farmers starting new farms in order to expand or to scale up their businesses. Especially these 'pig barons', who are keeping thousands to tens of thousands of pigs already and who want to start big new 'Mega-farms', arouse resistance in the areas of their choice, all the more so because they contract real estate agents and lawyers to help them realise their plans. Entrepreneurs of other activities in the national pig complex are also supporters of the pink invasion. Producers of compound feed and builders of stables hope to expand their businesses. Scaling up improves their chances of survival in the increasingly internationalising competition within their branch of trade. Part of the arable farmers in Zeeland Flanders and Eastern Groningen are supporters of a pink invasion as well. They want to stop farming in view of their age and the lack of a successor and hope to make a good deal with immigrating pig farmers by selling their farm or land. So supporters of the pink invasion are especially those who have a personal interest in the expansion of the pig sector in these regions.

That there are many *opponents* to the pink invasion seems somewhat strange at first sight, as there are several arguments in favour of an expansion of the pig sector in Zeeland Flanders en Eastern Groningen. Arable farming, the principal agricultural branch, is declining and intensive animal husbandry can offer an alternative or a possibility to supplement farm income. Moreover, arable farming and intensive animal husbandry can make a good production combination, as they can interchange manure and animal feedstuffs. In this way, transport costs can be avoided, which is a farm-economic as well as a macro-economic advantage, and less artificial fertiliser is needed, which is beneficial for both the farmer and the natural environment. Finally, in these economically weak regions with their high unemployment rates, every expansion of economic activities seems to be welcome.

All these benefits are disputed, however. The first important argument to the contrary is that the immigrating farms have very little to offer to the region. Firstly, it is a selective migration of the less gainful production of pigs for slaughter, while virtually all the necessities for production (piglets, stables, animal feed) are supplied from outside the region. Nor are the animals slaughtered or processed here, and the main waste product is manure, a worthless substance in a country with a big surplus. Furthermore, many inhabitants of Zeeland Flanders and Eastern Groningen expect a deterioration of their environment by smell pollution and the nuisance of increasing transportation for the 'Mega-farms'. Additionally, a number of people fear that specific gainful economic activities that are considered more important for the development of the regional economy than agriculture can be frustrated by the arrival of the pig farms. Smell pollution is not appealing to tourists, recreation seekers or well-to-do new inhabitants like old age pensioners. Others expect that the natural environment will be affected by an increase of intensive animal husbandry, or they have ethical objections to this form of production, factory farming. On the other hand, a moderate expansion of intensive animal husbandry is widely accepted, for instance as a supplementary branch for arable farmers, although there seems to be a general preference for attracting other agricultural specialisations, particularly horticulture and dairying. In support of this, one can point to the advertisements in horticultural journals by these same provinces and municipalities, in which they try to attract market gardeners and by the fact that the immigration of dairy farmers arouses no opposition. There must have been a substantial influx, as the number of dairy cows in Eastern Groningen has remained nearly the same over the past 18 years while dairy cattle in the Netherlands decreased by one quarter.

The objections listed above make clear who the main groups are that oppose the pink invasion: concerned citizens and their anti-pink invasion organisations, organisations for the protection of the natural environment and municipal and provincial authorities. But there is one more and interesting opposition group: arable farmers who fear that the immigrating pig farmers will employ all the potential space for manure disposal on the land, closing off animal husbandry as an alternative for them. In Zeeland

Flanders and Eastern Groningen, the number of pigs not only increases as a consequence of the migration of pig farms from the concentration areas, the *'hard-pink invasion'*, we might say. A number of regional farmers also switched to pig farming or started pig raising as an additional branch on their farm. This form of expansion of the pig sector evokes no resistance. On the contrary, opponents of the 'hard-pink invasion' applaud this *'soft-pink invasion'* as a sound initiative on the farm level that can have positive consequences for the region. This attitude can be an indication that, in choosing arguments for or against the 'hard-pink invasion', factors like regional initiative (endogenous development) and regional identity plays a role as well, apart from aversion against its massiveness and its threat of the environment. The generally positive judgement of the 'white invasion' (of poultry), which nearly always concerns an extra branch of an existing farm, also points in that direction. Perhaps the choice for consent or resistance may also depend upon the degree of commitment of the external actors involved. In contrast to many owners of the Mega-pig-farms, the farmers who start a farm in horticulture or dairying in the region exploit the farms themselves and settle down here.

Conclusions

In the above analysis, we have adopted a scale perspective on regional dynamics. The geographical scale of the social activities involved in regional development, and particularly the 're-scaling' of these activities, has been our focal point. This perspective highlights the following aspects of the migration of pig farmers in the Netherlands:

- The primary cause of the pink invasion is the strong increase of the scale of Dutch agriculture. Internationalisation and globalization not only brought about the expansion of the pig sector but also induced the stagnation of arable farming, and thus led to the emergence of strong and weak rural regions. The pink invasion, therefore, may be considered as an example of 'time-space compression' (Harvey, 1989): processes at high levels of geographical scale strongly influence developments at lower levels.
- Besides an increase of scale (internationalisation and globalization), a decrease of scale (regionalisation and localisation) also occurs in Dutch agriculture. Arable farming in peripheral regions of the Netherlands is increasingly grounded on regional and local impulses and resources (e.g., the physical environment, the labour market, accumulated local knowledge, and local patterns for linking

production to consumption). This is partly in response to the pink invasion into those regions.
- Despite the relatively marginal economies of Zeeland Flanders and Eastern Groningen, the two regions strongly resist the pink invasion. A number of small-scale regional alternatives to large-scale pig farming are being developed (e.g., agro-tourism and recreation, agricultural nature conservation, and organic farming). National regulation now also puts a halt to unlimited migration of pig farms. It seems that the globalization of agriculture cannot be regarded as an overriding process. Even peripheral regions are able to develop endogenous alternatives to exogenous impulses.

Regional identities are an important factor, notwithstanding our 'global age' (Albrow, 1996). The strong resistance to the 'hard-pink invasion' (pig farmers from Noord-Brabant) into Zeeland Flanders and Eastern Groningen is in sharp contrast with the support of a 'soft-pink invasion' (small-scale, second-branch pig breeding by farmers from the regions themselves). Apparently, pig breeding as such is not the main issue; more important are the scale of farming, its regional origin and the commitment of the pig farmers to the regions.

Finally, it may be noted that the pink invasion provides empirical evidence for a new direction in agrarian location theory. Traditionally, this theory was based on the choice of *production*, i.e., the kind of production on a farm on a particular location. The migration of Dutch pig farmers, however, suggests that in case of intensive animal husbandry, the choice of *location* should rather be the starting-point: the location for a farm with a particular sort of production. The same development can be observed for the Dutch glasshouse farms moving out of the overcrowded West, and for dairy farms generally relocating to the same arable farming regions as the pig farms. Agrarian location theory thus approximates to location theory for manufacturing and service industries. It might, indeed, be a 'myth that (globalized) agriculture differs significantly from other activities' (Le Heron, 1993, p.189). The overcoming of barriers between rural and urban disciplines would be a favourable consequence of globalization.

References

Agrarisch Dagblad (Newspaper for the Agrosector), 18 July 1998.
Albrow, M. (1996), *The Global Age: State and Society beyond Modernity*, Polity Press, Cambridge.

Boerderij (Agricultural Weekly), 3 March 1998, 30 June 1998.
CBS (yearly), *Landbouwtelling* (Agricultural census), CBS, Den Haag.
Gaasbeek, A.F. van (*et al.*) (1993), *Competitiveness in the Pig Industry*, LEI-DLO and Rabobank, Den Haag & Utrecht.
Harvey, D. (1989), *The Condition of Post-Modernity: An Enquiry into the Origins of Cultural Change*, Blackwell, Oxford.
Le Heron, R. (1993), *Globalized Agriculture: Political Choice*, Pergamon Press, Oxford.
LEI-DLO (various years), *Landbouw-economisch bericht*, LEI-DLO, Den Haag.
LEI-CBS (various years), *Land- en tuinbouwcijfers*, LEI-CBS, Den Haag.
Maas, J.H.M. (1994), *De Nederlandse agrosector: geografie en dynamiek*, Van Gorcum, Assen.
Maas, J.H.M. and Segrelles Serrano, J.A. (1996), Integración, cooperación y proyección exterior del sector cárnico-ganadero holandés, *Revista Española de Economía Agraria*, Vol. 178, No. 4.
Maas, J.H.M. and Segrelles Serrano, J.A. (1997), South and North in the European Union: The Livestock-Meat Sectors of Spain and the Netherlands, in T. Van Naerssen, M. Rutten and A. Zoomers (eds.), *The Diversity of Development: Essays in Honour of Jan Kleinpenning*, Van Gorcum, Assen.
Olijve, H. (1998), *Roze of groen? Een onderzoek naar regionale ontwikkeling in Zeeuws-Vlaanderen en Oost-Groningen*, thesis, University of Nijmegen.
Ploeg, J.D. van der and Long, A. (eds.) (1994), *Born from Within: Practice and Perspectives of Endogenous Rural Development*, Van Gorcum, Assen.

18 The changing agrarian structure of Israel's southern periphery

DAVID GROSSMAN AND HANNA MOSHAYOV

Introduction

In 1986, strong measures were taken by the Israeli Government to suppress the hyperinflation. The success of these measures sent the Israeli rural sector into a spiral of crises, which resulted in a real revolution. The main trigger was the high interest rate whose impact was especially hard on the farmers and on their co-operatives. It led to the dissolution of many farm co-operatives and eventually to the relaxation of many rules and regulations associated with the two major rural settlement forms: the moshav and the kibbutz. Co-operation in these settlements became optional rather than compulsory, as before. The strict communally managed mutual guarantee system was also abandoned in most moshavim.

Another revolutionary process was caused by the influx, since 1989, of about 700,000 Jews from the former Soviet Union and Ethiopia. The pressing demand to house and employ them was largely accomplished at the expense of the rural sector. The turnaround was expressed most clearly by the 'temporary' relaxation of one of the major national tenets – held by all former Israeli administrations – the protection of agricultural land. The accelerated process of turning agricultural resources into real estate fed on additional forces, which weakened the position of the agricultural sector. Eventually, many kibbutzim or moshavim were even encouraged to give up their tenurial rights (most rural land is National Domain), as a payment 'in kind' for settling their debts. This resulted in the creation of a class of 'Real Estate Settlements' located mostly in the core zone, particularly in the Tel-Aviv area. The marginal areas, where market prices for land are low, could hardly benefit from this policy, even though the government offered to pay higher prices for the land it took away.

This trend can also be seen in the spreading practice (often illegal) of using the land, originally leased for farming, for irregular and sometime illegal purposes. Studies have already shown that the principal pressure for building illegal storage structures is felt on land in the core areas, especially near industrial areas (Grossman, 1993; Sofer and Gal, 1995). The Negev is still largely free of these ugly structures. This is one of the few advantages of being peripheral. Even this blessing may not last very long, however.

The question is whether the process of change has been associated with marginalization of the more marginal areas? Has it widened the gap between the core and the periphery, or just changed its character? The present data indicate that agriculture had to retreat to the periphery, and this was accompanied by many social, economic and technological changes. This led to a reorientation of some national resources. Has this also resulted in an 'agrarian reform' especially in the marginal zones?

The purpose of this paper is to establish the nature of the emerging socio-economic patterns, to document and explain their impact on the land tenure system, and to assess the likely long-term outcome of the recent processes. This paper seeks to outline the benefits and the liabilities of the new revolution. The major issues to be considered are the following:

1. The use of unconventional (or secondary) water-supply, mainly from treated urban-industrial effluents;
2. The economic contribution to the national economy;
3. The social and agrarian impact of the transformation;
4. The Agricultural Potential of the Negev Periphery.

The contraction of the area available for agriculture in the country's core zone has already shifted part of the farming activity to the Israeli periphery. Most of this spatial transformation has taken place in the Northern Negev. The dry climate of this marginal area, whose annual precipitation is 200-300 mm per annum, accounted for the rather low population density and the patchy, unstable forms of its farming. It is blessed, however, with wide expanses of land, much of it covered with fertile loess soils. The low density of settlement has some advantages, however. Farmland is still available, and there is little competition from urban-industrial users.

Since the completion of water pipeline from the fountains of the Yarkon, which was linked, in 1964, to the National Water Carrier, the Negev has gone through a veritable transformation. However, water for farming continued to be a major constraint. The present new trends, associated with the demise of agriculture in the core, may result in a second revolution, which will drastically alter the landscape, the pattern of agricultural land use, and the population.

Unconventional water supplies - potential for the Negev agriculture

The key to the development potential of the Western Negev is a project which collects the effluents of all the towns of metropolitan Tel-Aviv, i.e., from an area inhabited by about 2 million people, filters them and treats them in a special plant south of the city. The project, called Shafdan, has been in operation since 1977, but ten years later it was expanded and the treatment methods were gradually improved. It now brings together additional waste water from a plant located a few kilometres to the east, and the combined treating capacity has been enlarged to about 115 million c^3.

After this treatment in the plant, the water is infiltrated into the underground aquifer of the coastal plain, and pumped again for transferring to the Negev. The methods employed in this system (Soil Aquifer Treatment) result in treated water, whose salt content is 290 ppm, compared to as high as 400 ppm. of the northern Haifa-Hadera project (Kanarek and Michail, 1996; Haaretz, 21.7.98). The chemical analysis indicates, in fact, that the Shafdan water is almost free of pollutant. In contrast with most other methods of waste water purification projects, its water can be used for irrigating practically all food plants, including vegetables. The water is fit even for direct human consumption, but as a precautionary measure, its use as potable water is prohibited.

The Negev receives less purified effluent water, whose use for vegetable growing is prohibited, from additional sources. Current projects, partly completed, are designed to transfer the wastes of Jerusalem. The waste waters from parts of the Tel Aviv suburbs which are still not served by the Shafdan will eventually be connected to the existing network, and various other urban areas throughout the country are expected to be served. Some of the existing projects, particularly in the Haifa zone, are expected to be diverted to the northern margins.

Altogether, the present amount of waste water which is used in the Negev amounts to some 130 million c^3. This figure is expected to rise to about 200 million by 2010. The ultimate amount of reclaimed waste water may eventually reach about 600 million c^3, close to one third of Israel's total supply of fresh water.

The use of reclaimed waste waters makes it possible to increase the extent of cultivated land, and accounts for the recent intensification of agricultural practices. This has resulted in raising the productivity of the land, and enables the freeing of fresh water for household use in the country's core where the population has increased. This clear-cut distinction between the main classes of users will eventually lead to the total dependence of farming on reclaimed water. This can be regarded as a rational means of allocating scarce resources, and as a clear contribution to their efficient use (Weinstein, 1994).

Irrigation by purified water necessitates light soils having high porosity. This makes sandy soils ideal for these purposes. The sands of Halutza in the southern margin of the Western Negev has become an important resource for the use of these waters. The sands have the added advantage of retaining warmth better than other soil types. Furthermore, since they are almost completely sterile, they respond well to controlled management. Fertilisation can be accurately applied according to specific plant needs. The production level is, thus, also controlled.

The warm winter of the desert has advantages also elsewhere. The main advantage is the out of season marketing of early maturing crops. High temperatures allows the implementation of agricultural methods unsuited for the centre of the country,. This includes savings on heating the large (up to 4 ha each) plastic covered structures, which are now commonplace in the Western Negev. In most other zones of Israel (particularly the Jezreel Valley) heating is mostly required during the coldest nights of the winter. Expenditures on pest control are also relatively low because of the climatic conditions. This benefit is directly related to the main disability of the Negev: its dry climate. These regional advantages gave rise to expectations for development prior to the implementation of the treated wastewater. It was correctly argued at the initial stage of the development that under desert conditions only by means of sophisticated systems, unconventional initiatives, and large investments, along with careful planning of the marketing phase, could specialised crops succeed. High investment levels were, therefore, required (Amiran, 1977).

The implementation of the transformations in land use and water utilisation has been facilitated by research and development projects assisted by the Office of the Head Scientist of the Ministry of Agriculture. Agriculture in the Western Negev is now expanding. Since 1992, about 3,000 hectares of orchards have been planted. Former rainfed areas have come under irrigation. Field crops have been replaced by more profitable crops, such as citrus, tropical fruit, vegetables and flowers (Figure 18.1). The last two are grown under plastic cover (Kedar, *et al.*, 1994; The use of glass, common in the greenhouses in cold European regions, is not necessary in Israel). Intensification is on the rise. This necessitates greater financial investments. This usually generates a 'chain reaction' i.e., a desire to enlarge the farm units in order to maximise returns by taking advantage of scale economies.

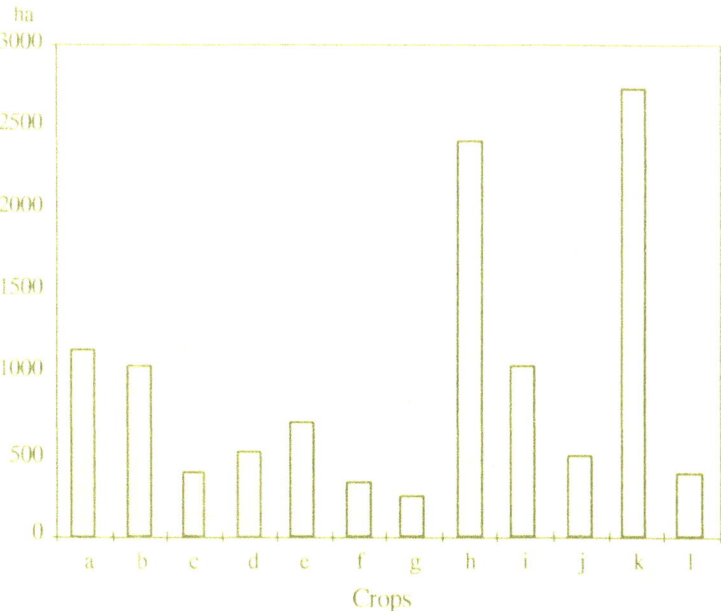

a) tomato, under plastic, b) tomato, under net, c) tomato, cherry, d) pepper - under net, e) pepper - open, f) other vegetables, g) roses, h) other flowers, i) mango, j) sabra, k) citrus, l) other fruits

Figure 18.1 Land use Eshkol Regional Council, 1998 season

Source: Eshkol's Agricultural Commission.

The process, however, is far from uniform. Very close to the Eshkol zone, there are other regional councils where farming is very extensively practised, if at all. The causes for this intra-marginal polarisation is not fully understood. We hypothesise that its occurrence is explained largely by cultural factors.

The expansion of the area under cultivation in the Western Negev necessitates an increase in the amount of water available for agriculture. This can be performed by diverting water released from the neglected or abandoned agricultural lands of the core, where building on former farmland is now a normal practice. It is estimated, thus, that about 10,000 hectares of citrus groves have been cut-down or discarded in the Sharon Plain (Figure 18.2).

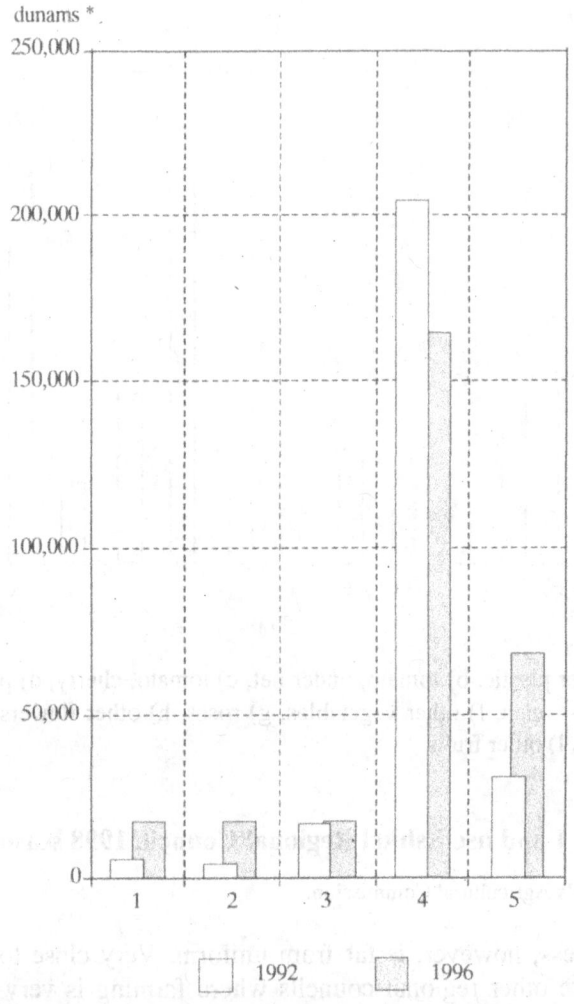

* one dunam equals 0.1 ha, 1) Upper Galilee, 2) Lower and Western Galilee, 3) Jordan and Sea of Galilee Valleys, 4) Sharon, 5) South.

Figure 18.2   Main citrus growing zones in Israel, 1992 and 1996, in dunams

Source: Gadish, 1994.

Since the reclaimed effluents include salts and other chemical substances, they are not suitable, in fact, for agricultural use in the core zone where they may contaminate the fresh water aquifers. It is important to note that, because of the high porosity of the Sharon's sandy soils, the chemicals present in the treated effluents do not permit its use there. The water is much less harmful, however, when used on the sandy loess of the Western Negev (Fein and Dan, 1997). The reclaimed effluents contain, in fact, some chemical fertilisers which add to their attraction for agricultural use. This means that the agricultural consumption of this water in the more sandy loess soils of the Negev does not compete with the core region of Israel. Citrus yield in the Western Negev are, in fact, higher than those of the Sharon, and compare favourably with the latter also in their input/output ratio. Studies demonstrate, however, that, because of the high cost of water, cultivated crops have to be carefully selected (Brimberg *et al.*, 1997).

Even though the most important future increase in the Negev's water consumption depends on the use of recycled effluent water, some of the added demand may be met by expanding or improving the utilisation of local water resources. The total water consumption for agricultural purposes in the country as a whole is not expected to rise. In fact, the plan calls for reducing the share of agriculture in the fresh water utilisation (Yogev, 1995; Cohen-Kedmon, 1998). Continued development in the twenty-first century will depend mainly on the supply of purified effluents, as well as on a small amount of desalinised brackish water.

This trend has already started. In 1990 the Western Negev settlements relinquished their rights to 100 million m3 of fresh water, which were transferred to the growing urban sector of the area. They received, in exchange, purified effluent water (Tamari, 1995). The quotas of fresh water for agricultural consumption are also reduced in dry years (Lipschitz, 1996).

Economic consideration and spatial costs

The cost of the pumping and transporting the water has been a controversial issue in the country. The critics agree, mostly, that agriculture is no longer profitable. It is argued that, even on the national level, the highly subsidised water supply, which is sold to the rural settlements at a fraction (15%) of the production cost (pumping and distribution), amounts to 'exporting water' rather than farm products. The share of agriculture in the economy of

Israel is constantly declining, and employment in agriculture is now about 2% of the labour force, similar to that of the USA. The Jewish population of Israel refrains from agricultural employment, and most of the rural dwellers are either 'managers' or off-farm workers. Farm labour is provided mainly by imported labour from Thailand, or from Gaza and the West Bank. In 1997, there were 1628 Thai workers in Eshkol Regional council. Of this total, only 90 were employed in the Kibbutzim (communal villages).

Those who favour agricultural development point to the vital role played by farming in any national economy and to its security importance during emergency conditions. There are also emotional and ideological considerations, which can be viewed as 'residual values' of the original Social-Zionist ideology. However, the common argument is that agriculture is still the most effective means for establishing national control over the land, and plays an important role as in maintaining open spaces. It is also important as a landscaping devise. Subsidies to agriculture have been progressively reduced, and are now much lower than those of most European countries or Japan. California farmers pay only 5% of the cost of the water they use, about a third of the share paid by the Israeli farmer.

Those who favour agricultural development point to the vital role played by farming in any national economy and to its security importance during emergency conditions. There are also emotional and ideological considerations, which can be viewed as 'residual values' of the original Social-Zionist ideology. However, the common argument is that agriculture is still the most effective means for establishing national control over the land, and plays an important role as in maintaining open spaces. It is also important as a landscaping devise. Subsidies to agriculture have been progressively reduced, and are now much lower than those of most European countries or Japan. California farmers pay only 5% of the cost of the water they use, about a third of the share paid by the Israeli farmer.

The controversy over the new Negev projects depends on the sides taken on the national level. The cost of purifying, filtering, pumping and transporting water to the Negev is seen as totally unjustified by the critics. 'The user should pay the total cost'. However, the economy of waste water use must take into account that the Shafdan and similar projects have been set-up primarily to solve the growing ecological problem created by urban sewerage. The only alternative to purification is the continued pollution of the Mediterranean. Agricultural use is clearly a better and more desirable solution. Economically, therefore, it is quite just to expect that urban dwellers, the producers of the waste, pay the cost of sewerage treatment (Haruvi et al., 1996).

The only issue which directly involves the Negev project is geographical, i.e., the cost of transferring the water to the Negev. Our previous review of the alternative potential uses, illustrates, however, that the feasibility of use of the effluents in the core zone is minor indeed. It should be stressed here that the policy of the water company, Mekorot, is to even out the water prices without regard to distances. This 'price discrimination' against core users (see Hoover, 1948) has been an important contribution to Israel's policy of spreading population and evening out regional differences. Whatever the actual merit of this policy, there is little doubt that the new Negev project can be justified even on purely economic grounds.

Aside from the price of water itself, the use of treated effluents requires added investments by the farmers. This is by no means a small amount. They have to obtain special equipment and filtration plants. Additional extra costs include the large amount of water used for cleaning the filters and washing the land and the equipment. The land and the vegetation quality have to be repeatedly checked, the crops carefully selected, and damages in the operating system, either natural or man made, must be carefully repaired. The costs involved in these operations have to be carried by the farmers themselves.

Social implications and agrarian impacts of the recent transformation

Technological development which brings about an increase in yields, a fall in price and a decline in profits will not suffer small farming units. Therefore, agrarian changes will take place modifying tenure systems and bringing about the creation of large land units by different bodies. Farmers who are in debt and who are unable to find sources of money or credit from the banking system have to trade off their lands and production quotas which till now were their sources of income to large companies which have decided to invest in agriculture for commercial gain. Other farmers take over the unused property of their neighbours, or rent areas in adjoining moshavim thus forming large units in order to put modern technologies into effect and exploit superior crops and cheap foreign or Arab labour.

On the other hand, there remain settlers who have relinquished the lands given to them by the Israel Land Authority. An inequality is thus established between those who continue to be active agriculturists and those who have given up such undertakings because of the lack of profit and gain. This has brought about a revolution in Israeli agriculture which has become motivated by economic factors and has concentrated production factors in the hand of just a few, thus destroying the original ideal of 'equality' that was considered to be its base. If in other countries, especially those of the Third World, former farmers have tended to move to the cities, in Israel, like in developed countries, because of the close geographical proximity of the towns to the villages, and because of educational levels, there has not been an abandonment of the latter. People remain in the rural areas where life quality is considered to be superior, and gain their livelihood in the towns. The level of commuting increases. Levels of poverty do not change because the alternative livelihoods in the city and the opportunities there for increasing education allow the non-agricultural workers who continue to live in rural areas while they work in the cities to maintain good standards of income. Many of the social transformations dealt with here were founded in the agricultural crisis of the eighties and have spatial implications. This is especially true in the realm of the distribution of financial and land resources. The refusal of the banks to grant credit to farmers who have not paid off their debts, and that of the government to finance agricultural endeavours mainly hurt farmers living on the periphery. The new expansion of citrus groves and other plantations is based on new models of agrarian and organisational forms. Some of them are listed below:

1. Inter-kibbutz and inter-moshav co-operation, on pooling regional resources to take advantages of large-scale operations. The farm work is not necessarily carried out by the local population. A number of large agricultural companies, which have specialised in providing contracting services are now involved. They enter into long term agreements with the local farmers, according to which the land is fully managed by them for a period of fifteen years. The companies invest in seed and other inputs, and employ foreign workers (Gadish, 1994). This type of arrangement is found particularly among settlements which had accumulated large debts as a result of the 1986 crisis. One of the new projects called for planting 10,000 hectares of citrus orchards. It attempted to withdraw land use rights from the moshavniks who neglected their farms, by offering them monetary compensation. The purpose of this policy was to pool land resources for planting citrus orchards. It failed, however, to find many subscribers (Gadish and Zohar, 1994).
2. For similar reasons kibbutzim have banded together to plant and work large orchards. Such practices cause the abandonment of the principle of self-work and replace it by cheap 'guest labour'.
3. Farmers who have financial resources can now utilise them to sublet properties from neighbours or neighbouring moshavim. This practice results in breaking down the former equitable resource distribution rules. There are

already some moshavniks who operate properties which extend over many hectares.
4. Large marketing companies such as 'Tnuport' and 'Mehadrin' have begun to invest in planting citrus orchards in the Negev. These concerns have taken the place of the government that, until recently, was responsible for providing vital means of production. The private concern now supplies even water by purchasing the water rights of individual farmers who are unable to obtain capital for the investing in new agricultural projects. Such practices were formerly illegal. (An example is a newspaper advertisement of known business concerns offering to acquire the rights of usage of twenty million cubic meters. of expected purified water of the Jerusalem sewerage.) Contracting companies are involved in all facets of agricultural production, and may have already taken over possession (but not ownership!) of much of the Negev's land.
5. Large private or public farms, hitherto rare in Israel , are becoming widespread in the Northern Negev. The growing use of purified sewage water, which makes it possible to expand agriculture, is partly responsible for this change. Some of the new private farms extend over hundreds of hectares and are used mainly for pasture. They raise a variety of animal (one of them raises ostriches), though they usually also grow some crops. They resemble, in fact, American ranches. In Israel, where most of the farmland is held by small-scale family farmers or by communal groups, such ranches are still unconventional. Most of these ranches are located in the drier parts of the Negev and in rocky areas, but several of them are in the Western Negev.

One of the early manifestation of the undercurrents which eroded the principles of the Israeli moshav was the establishment, as early as 1962, of a company called 'Hevrat Moshavei Hanegev'. It was set up to work the land of some 34 moshavim of the Merhavim, Bnei Shimon, 'Azata and Sha'ar Hanegev regional councils. This public company employs Jewish workers and managers along with foreign and Bedouin workers. This system was officially promoted despite the fact that at that time it negated the basic rule that a moshav member should not use hired labour on his land. This rule is now mostly ignored. The very fact that such a company was established is an expression of two sets of conditions:

a) The Negev is one of the only places in Israel were large tracts of land are available for extensive cultivation;
b) Most of the Negev's settlers were immigrants from Arab countries who were not brought up under the socialist ideological influence. They also had

little training in agriculture. Most of them are still poor farmers and prefer non-farm employment.

Summary and conclusions

The experience of the Western Negev demonstrates that technology is capable of altering the status of marginal zones, and may even have the potential of turning the periphery into an economic core. However, in practice this is not necessarily the case. Furthermore, even if the regional income rises, the social outcome of this transformation is debatable. The comparison with the Sharon Coastal Plain (north of Tel Aviv) is instructive. Until about 1920 the Sharon was a marginal zone. It was sparsely populated and had few permanent agricultural settlements. The citrus industry changed it, and it is now the densest rural area of Israel. Geographical location has obviously played a vital role in this transformation. However, the social values and the work ethic were also important. The change was largely due to small-holders who worked and managed their own farms, and were strongly motivated to socialist ideology. In the present Western Negev, most family farms are family managed, but not family worked. Some of the few who still stick to the older value system view these developments with strong concern. They see the present trend as leading to the creation of latifundia, that is, the gradual concentration of land resources in the hands of a small number of owners who depend on the employment of foreign (mostly Thais) low-paid workers (Zamir, interviewed 4.5.98). One of the other critics of this change, which is supported by official policy, puts it this way: 'from farming practices where a large number of cultivators work small parcels of land, we have changed to a system where fewer farmers produce large quantities' (Galily, interviewed, 4.5.98).

Technological and economic change is linked, thus, to the transformation of the social, the agrarian, and the value systems. These systems are linked to each other in a way which is not easily predictable. The changes do not occur only in the marginal zone. They take place all over the country. Locational factors make the Negev conditions stand out as candidates for agrarian change. Core locations, on the other hand experience the demise of agriculture, and the process of real-estatization, which results in the conversion of farmland into residential suburbs and industrial parks. Thus, the difference between core and periphery is not obliterated. It just changes form. In fact, the easing the restrictions on using agricultural land has widened the gulf between settlements of the core and the periphery.

References

Amiran, D.H.K. (1977), 'Unconventional Development in the Negev', *Nofim*, no. 9-10, (Hebrew).
Brimberg, J., Oren, G. and Mehrez, A. (1997), 'An Operational Model for Utilizing Water Resources of Varying Qualities in an Agricultural Enterprise', *Geography Research Forum*, vol. 17, pp.67-77.
Cohen-Kedmon, M. (1998), 'Creative Activity in the Family Farm, Spatial, Physical and Organizational Aspects', Committee Report, presented to the Israeli Ministry of Agriculture (Hebrew).
Fein, P. and Dan, J. (1997), 'Soil Considerations in Combining Effluents Sludge and Irrigation in Agriculture Use', *Water and Irrigation*, no. 369, pp.30-40 (Hebrew).
Gadish, J. (1994), 'Comprehensive Strategy for the Citrus Industry', Final Report to the P.O.C. Research, Citrus Marketing Board (Hebrew).
Gadish, J. and Zohar, M. (1994), 'The Citrus Industry, Future and Present, Survey of Varieties and Regions, and Development Conditions', Invited by the Citrus Marketing Board (Hebrew).
Grossman, D. (1993), 'Non-Agricultural Penetration of the Moshav in the Eighties - Effects on Land Use', *Karka* (Land), vol. 37, pp.44-59 (Hebrew and English summary).
Haaretz (Daily Paper, Hebrew and English), 21.7.98.
Haruvi, N., Fein, P. and Scheinberg, I. (1996), 'Economic and Holistic Considerations Concerning the Reclaimion of Wastewater for Agricultural use', *Water and Irrigation*, no. 353, pp.17-21 (Hebrew).
Hoover, E.M. (1948), *The Location of Economic Activity*, McGraw Hill, New York.
Kanarek, A. and Michail, M. (1996), 'Groundwater Recharge of Municipal Wastewater - The Gush Dan Reclamation Project', *Water and Irrigation*, no. 356, pp. 13-17.
Kedar, P., Oren, A. and Wagger, M., (1994), Growth Structures in Gush Qatif, *Technology and Manpower Alternatives*, Development Study Centre, Rehovot.
Lipschitz, E. (1996), 'Waters Too Precious for Agriculture', *Water and Irrigation*, no. 359 (Hebrew).
Sofer, M. and Gal, G. (1995), 'Environmental Effects of the penetration of Non-Agricultural Industries into the Moshav', *Horizons in Geography*, 42-43, pp. 39-50 (Hebrew with English Abstract).
Tamari, J. (1995), 'Of Management and Authoritarianism in the Water Economy', *Water and Irrigation*, no. 349, pp.4-5.
Weinstein, Z. (1994), *Supplying Effluents to the Negev's Citrus Orchards, Presented to Citrus Marketing Board*, Z. Weinstein Engineers and Advisers, Tel-Aviv.
Yogev, D. (1995), 'Proposal to Enlarge the SHAFDAN Project to 200 million cubic meters, and its incorporation into the Agricultural Production System of southern Israel', *Water and irrigation*, no. 339, pp.36-43.

19 Coping with business crisis in the Russian North: Strategies against turmoil

MARKKU TYKKYLÄINEN

Introduction

Industrial output has declined in the Russian North throughout this decade. In parallel, economic modernization has not succeeded as anticipated and industrial capital is ageing due to lack of investments (e.g. Tykkyläinen and Jussila, 1998). This paper elaborates upon the reasons for this turmoil by analysing the development of two companies, a nickel conglomerate in Murmansk Oblast and a former ski, furniture and parquet company in the Karelian Republic. The study analyses typical economic crises in resource communities in the Russian North. In keeping with the definitions employed by official Russian Statistics, the areas referred to as the North are Murmansk Oblast, the Karelian Republic, Arkhangelsk Oblast, Vologda Oblast and the Komi Republic.

The paper aims at constructing a theoretical framework for conceptualising the disintegration of 'production complexes' in the Russian North. This study is an independent piece of research and forms part of two wider research projects which aim to analyse local development strategies in several communities in Russia and Hungary (see Tykkyläinen *et al.*, 1998; Tykkyläinen and Rautio, 1998; Tykkyläinen, 1998a). The empirical part of the study relies on research of business and livelihood in the small suburban locality of Helylä, in the village of Värtsilä in the Karelian Republic and the mining towns of Zapolyarnyj and Nikel in Murmansk Oblast on the Kola Peninsula (Tykkyläinen, 1998b; Tykkyläinen and Rautio, 1998).

Economic decline and transition

Socio-economic transition has proved to be a long-term process in many remote places of Eastern Europe and especially of Russia. Individuals and various local communities have encountered difficult adjustment processes with the worst situation being in the countries where production has declined the most. The average change of GNP per capita between 1985 and 1995 was -5.1% in Russia (World Bank, 1997, p.215). The economic regression has continued through to the late 1990s. In 1997, the GNP of Russia recovered slightly but declined in the other years: -3.5% in 1996 and -1.1% during the first seven months of 1998 (Bank of Finland, 1998). The competitiveness of Russian industry has been low in the latter half of the 1990s which was one of the factors leading to the floating of the Russian rouble and its recurrent devaluations from 17 August 1998 onwards. Karelia and the Murmansk Oblast, the regions where field research for this paper was undertaken, have suffered from the decline of resource-based industries, and modernization is urgently needed for most of the companies in these regions.

Russia has witnessed the evolution of a unique method of the dissolution of socialism among the countries of the former socialist bloc. In spite of such profound changes, Russian authorities have nonetheless been unable to modernise organisations and institutions according to the requirements of a market economy (see e.g., Tykkyläinen and Jussila, 1998). Additionally, Russian industrial towns are much more isolated from western markets, finances and technologies than East-Central European ones, further lending to the unique predicament of Russia.

Former results: actors of development

Despite a sizeable bureaucracy it would be an exaggeration to say that Russian authorities currently plan and implement effective, sustainable local development strategies. Development is sporadic and local development consists of rather scattered struggles against turmoil by individuals or collective groups rather than any plausible development strategies.

The term survival strategy, or coping strategy, has been introduced in different contexts in recent research. In the most abstract sense, such a strategy is considered to express the reproduction of capital-labour relations. A survival strategy also has more concrete meanings related to the increase of informal work and the livelihood of low-income groups (Mingione, 1991; Meert *et al.*, 1997). On the whole, a coping strategy consists of reactions which individuals, companies and authorities adopt in the face of a local economic crisis (Nygren and Karlsson, 1992). In transitional countries, the restructuring of the local economy in small communities, such as towns, has led to drastic attempts to satisfy basic needs by inventing new means of livelihood, resulting in a survival strategy for individuals and families (Tykkyläinen, 1998a; Voronkov, 1995). At the practical level, individuals and communities must develop new strategies to restore incomes through, for instance, innovating new products, increasing the efficiency of work, establishing new enterprises, increasing of enterprising work and adopting various forms of informal or casual activities.

There are numerous case studies which highlight how entire localities have transformed their economic and social structures during crises, and how different strategies of survival or development were implemented at a community, usually town, level (Neil *et al.*, 1992; Neil and Tykkyläinen, 1998a). These former studies revealed that the actors reacting to the pressures of restructuring are not great in number but rather a handful of people from both in and outside of a community (Neil and Tykkyläinen, 1998a). These actors are comprised of individuals and newly formed groups and *ad hoc* organisations. Restructuring usually supersedes the borders of a single community and brings together new resources (skill, funds, etc.). However, it is not realistic to expect that a traditional community – a local authority or local residents – will operate as a collective and coherent organisation in the restructuring phase. Indeed, the consequence of the heavy pressure to restructure is usually disorder rather than increased cohesion.

Hypothetically, Russian areas rich in natural resources form arenas for various efficiency-seeking processes, which would lead to the active search for new development and business strategies in order to utilise natural resources and existing social capital. In Russia, society is not organised in such a way so as to permit the implementation of efficient development policy by regional and local authorities. Nor are the regional and local authorities authorised with the unambiguous power or equipped with the necessary financial resources for implementing such policy at present. The potential for development still depends on old societal structures and emerging capitalism.

Working communities, where the provision of services (infrastructure, housing, central heating, etc.) is associated with the operation of factories still predominate in many places in Russia, and local authorities play a minimal role in community service provision and development, notably in the Russian North. The issue of organising local development seems to be problematic. In outlying areas, local authorities do not have many revenue sources, and local people remain dependent on benefits provided by factories. In general, most people do not possess the ability and resources to promote development and, hence, are not prime actors in development (e.g. Tatarinskii, 1998).

As concluded from earlier research, individuals, small groups, entrepreneurs and investors are the real actors of restructuring a community. A community consists of different skills and capabilities, different occupations and generations and, of course, different individual values and attitudes. The reactions of individuals and *ad hoc* -groups to restructuring are direct, spontaneous and far-reaching. All this leads to the assertion that there are impromptu and unexpected actors in Russia reorganising economic systems.

The explanation of economic transition is elaborated in this paper by the investigation of the conditions of transition at a company level. An analysis of the cost structure of companies is used in explaining the decline and emergence of economic activities in contemporary Russia.

Theoretical framework - actors in Russia

Institutional context

Previous work undertaken with the restructuring of resource communities resulted in several conclusions. First, that there are general transition processes (such as globalisation, transformation of general socio-economic philosophies and re-organisation of international trade) forming the primary external factors steering development. Second, national political factors (such as privatisation, the introduction of market economy rules, etc.) challenge the industrial communities to improve their competitiveness (Neil and Tykkyläinen, 1998b; Tykkyläinen, 1998c). Third, there are differences in the future growth opportunities of industry according to industrial sectors. Fourth, local development processes are locally-specific, implying that they are based on the unique configuration of population, infrastructure, local economy, institutions and local cultures. These supra-local and local processes overlap and local survival strategies emerge in the interface of these processes. Fifth, individual actions, such as 'survival strategies' represent individual responses in a community (Tykkyläinen, 1998c).

In conditions of turmoil, development originates from pressure on communities to change. This means that various factors (such as shifts in demand, deregulation, etc.) exert pressures on villages, towns and rural areas to restructure and, in more general terms, foster the development of a community. Individuals and groups react to these pressures resulting in what may be called strategies against marginalization. People react to change in various ways such as resisting change, passively adapting to changes or attempting to be innovative (Nygren and Karlsson, 1992, pp.110-116). Innovative behaviour, in turn, lends to the 'development' of a community. Whatever the strategy is, the influence of the various factors (and the interaction of them with actors) produces different spatial outcomes (e.g. closures of plants or 'development').

The interplay of structure and agency – development

One of the traditional debates of relevance in explaining Russian socio-economic development is the relationship between structure and agency. That debate can be referred to as the discussion of the nature of human geography (Thrift, 1983). In its simplest, development is an interplay between human agency and economic structures (Figure 19.1). From the view of an empirical standpoint, development is an interplay of actors and institutions.

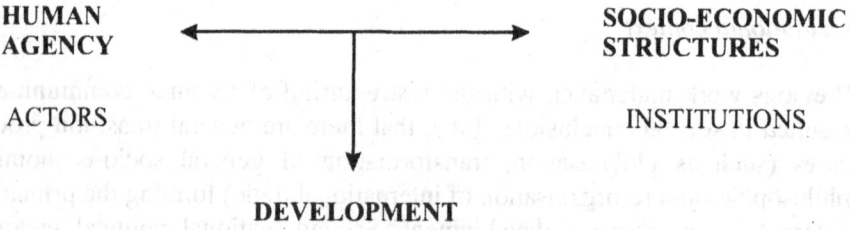

Figure 19.1 The interplay of human agency and structures

The role of human agency is fundamental in reactions to restructuration processes. Human activity shapes everyday livelihood and generates the economic viability of a community. Simultaneously, and from the viewpoint of geographical social theory, human activity is an evolutionary process which creates economic, spatial and social practices and, finally, structures. Human agency, comprised of a complex web of actors, is central in creating conditions and acquiring benefits from any economic transformation. Local development processes are learning processes with feedback.

Besides transition impacts on everyday livelihood in Russia, one outcome of the transition has been the emergence of a great number of reformed or nascent social structures. New social practices emerge and shape new institutions and structures, but only very slowly. Industrial communities are laboratories of ongoing restructuring in Russia.

The transition in the Russian North indicates that modernization is an intentional process (involving various actors) with certain socio-economic regularities (such as profit-seeking) and is affected by the past social system (i.e. institutions, organisations and economic structures). The past poor performance of industrial plants and the former division of labour, based on non-market pricing, have induced a profound but excruciatingly-slow restructuring of industry. Russia is developing its unique form of transition, currently consisting of a combination of laissez-faire, a more conventional market economy as well as modernised socialistic practices (Tykkyläinen and Jussila, 1998).

Modernization takes place in locations that are suitable for profit-making. Restructuring is a spatially-uneven process due to the legacy of the former spatio-economic structures of Russia's former command economy. Emerging capitalism selects locations on the grounds of profitability, and hence, constructive restructuring takes place in favourable 'pockets'. The pattern of uneven spatial development is discernible if one examines development in individual locales.

Conditions for development: two case studies

Radical changes

There are only few examples of successful business developments in the Russia North. The following two examples represent strategies by which some people attempt to accomplish development within the spheres of declining communities.

Helylä

A former ski, furniture and parquet company is located in Helylä (61°55'N 30°38'E), a few kilometres from the town centre of Sortavala (36,000 inhabitants). The company has been responsible for supplying the employment and utility needs of Helylä and its 4,000 people. The main plant was in Helylä, but the company also had a farm in the vicinity as well as a subsidiary plant. When the company operated at full capacity, the number of employees was 2,500. The company reduced its labour force in the 1990s, and in 1997, the company employed only 700. The company went bankrupt in the beginning of 1998 (Tykkyläinen, 1998b).

When the company collapsed, the social obligations of the Helylä community were transferred to governmental authorities. The discarding of social obligations made it possible for company managers to begin new economic activities without the burden of providing public utilities and services to the entire community (Tsaplin, 1998).

As a response to sudden employment, the most proactive former employees of the company focused on forming three business ventures in early 1998. First, a new company with premises in Sortavala was established to manufacture furniture. The operation of the furniture company was still in its infancy in February 1998, with the business strategy of the company being to produce furniture and kitchen fixtures. Second, there would be a continuation of the production of skis and ice-hockey sticks, but under the auspices of a new company. Third, a small-scale sawmill began to operate on the premises of the company's farm, commencing production in January 1998.

The sawmill machinery was located in the farm of the former company. The farm still had 50 cows and grassland fields of 160 hectares in 1998, but other parts of the farm were dedicated to the sawmill operation. In the winter of 1998, the sawmill operated in an old cattle house and employed 28 workers. The company planned to recruit more people when three-shift operation would begin and production would increase. The venture capital of the sawmill was put together by investors from Hungary, Lithuania and Russia with respective shares of 40/33/27. Sawn timber is exported to Budapest, Hungary by lorries. Roundwood is transported to the mill from as far away as 40 kilometres, and the company has an agreement for logging 10,000 cubic metres of roundwood per annum. Compared to many operations, this company is very small.

Zapolyarnyj and Nikel

Zapolyarnyj (69°26'N 30°52'E) and Nikel (69°24'N 30°14'E) are industrial towns near the Russian-Norwegian border in the Murmansk Region. The industrial base of these towns consists of nickel production operated by Pechenga Nickel Company: three open pits, underground operations, a mill producing nickel concentrate, a roaster plant producing pellets, a sulphuric acid plant and a smelter are the main production units in the 10x30 km mining area located in the arctic region. The mines and factories employed 9,000 in August 1998. Retirements and layoffs were seen as a valve for financial woes and 800 employees became redundant in 1997 and the labour force was further reduced by 1,100 in 1998.

Pechenga Nickel is part of Norilsk Nickel Inc, the subsidiary of the large multilocal mining company in Russia. The main worry of Pechenga Nickel is the lack of competitiveness and profitability. The current low price of nickel does not account for all the losses, with the primary reasons being the lack of advanced organisation and inefficiency of production. During research interviews, managers of Pechenga Company voiced concerns of insufficient resources for required modernization investments (Blatov, 1998; Kamkin, 1998). The company also needs investments to develop a new underground mine to ensure sufficient nickel ore for the processing works.

Co-operation has been planned with the Finnish company, Outokumpu Oyj, which is expected to provide modern technology and equipment. Pechenga Nickel has constructed two shafts (more than 1,000 m deep) and supplementary facilities, representing their inputs to the possible joint venture. If this joint venture does not materialise, nickel production in Zapolyarnyj and Nikel will cease (*ibid.*). The Zapolyarnyj and Nikel case study exemplifies modernization and restructuring attempted through cooperation with a foreign company.

Explanation: restart through crises

Transition has so far lasted almost ten years in Russia. Why are companies unable to develop their production? Below, the explanation is discussed in a more analytical fashion using some traditional concepts of economics.

The direct reason for unproductive economic performance seems to be obsolete products and technology in both case studies. Poor economic performance can be simply illustrated by shifts in technology, investments and demand in two countries. Take the hypothetical case where consumers increase or decrease the demand, say, for skis over the long run. Assume that the equilibrium of demand and supply in market economy country A is described as E in Figure 19.2, and non-market economy B can provide the same amount of products at the same price. The starting point is that the demand and supply curves, dd and ss, are located in the middle (Figure 19.2), and there are companies in both countries producing almost similar products at the same price.

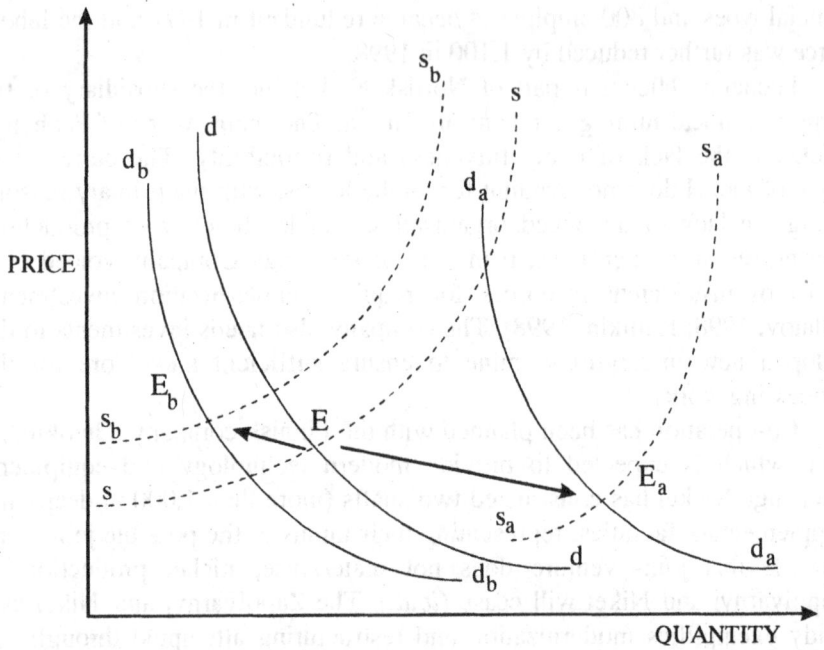

Demand curves ———— , Supply curves - - - - - . Subscript (a) refers to advancing market economies and (b) to Russia

Figure 19.2 Downward shift in costs and increasing demand in growth economies and upward shift in costs due to declining (mass) production and decreasing demand due to decreasing incomes in declining economies

The know-how gained in the production process and new innovations make possible the production of the next generation of skis for much less per unit than before in the developing market economy A. The demand increases because incomes increase and consumers feel that new skis are more appealing than older pairs. The demand and supply curves shift to the right in the case of country A as well as the long-run equilibrium of supply and demand from E to E_a (Figure 19.2).

In the nascent market economy, country B, high volume production is no longer possible and skis become outdated in comparison with skis produced elsewhere over time. Thus, the price of skis manufactured in country B rise due to rising costs and consumers are no longer equally eager as before to buy these skis. Hence, the gap between the competitiveness of these two countries widens in ski production. The shift to equilibrium E_b depicts the market development in Russia, the direction which is opposite to the shift towards equilibrium E_a which illustrates the situation where companies are innovative and develop their production, as is the case in country A (Figure 19.2).

But why do the companies in country B fail to develop their production? This can be explained by the concept of former inherited structures (cf. Tykkyläinen and Jussila, 1998). The legacy of the former economic practices is elucidated by analysing the development of the ski, furniture and parquet company which was in operation in Helylä in the 1990s (Figures 19.3 and 19.4).

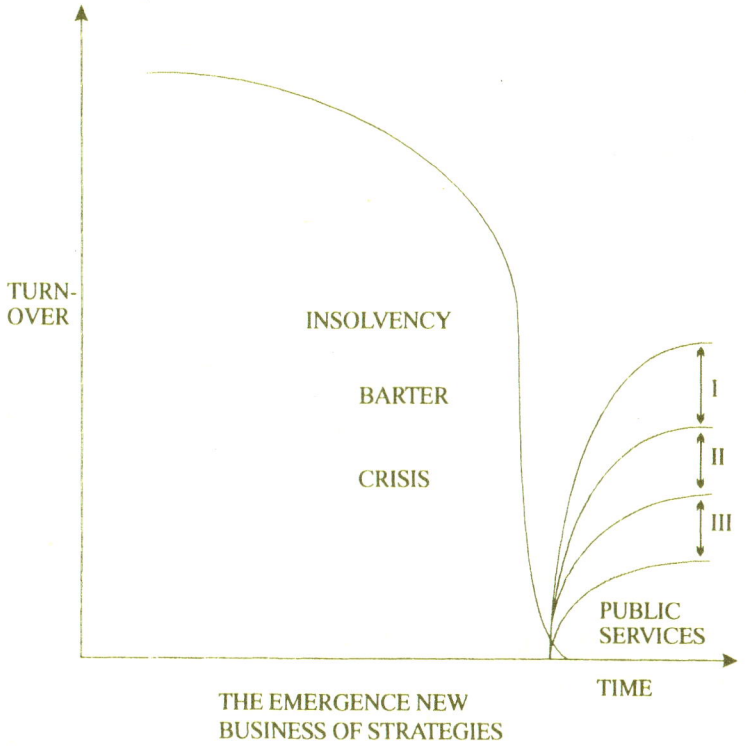

I, II, AND III DEPICT RECENTLY FOUNDED COMPANIES

Figure 19.3 The closure and restart of economic and social activities

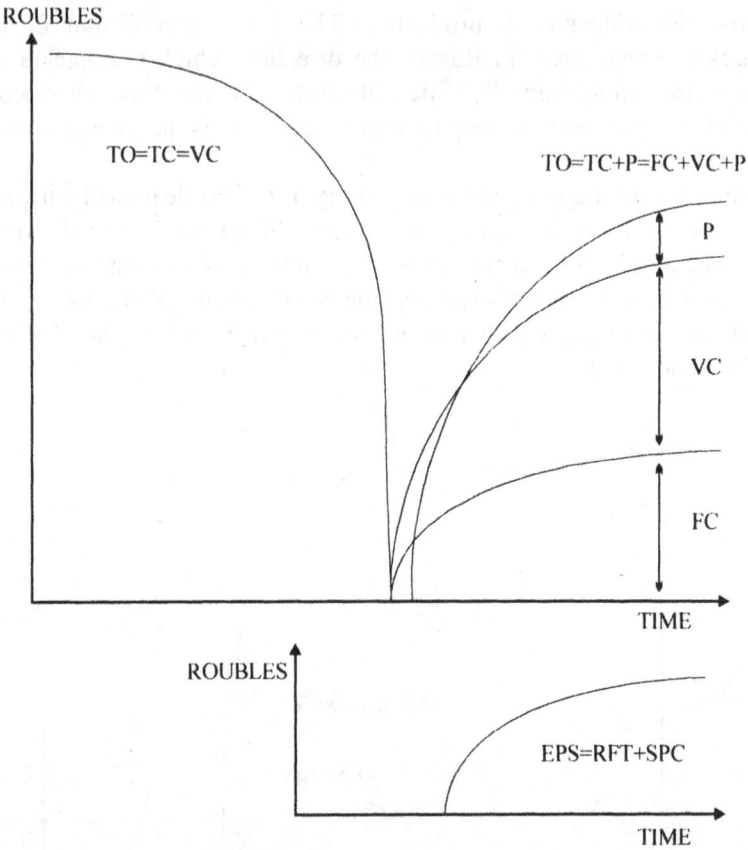

Figure 19.4   Revenues and components of cost in old and new companies

The curve on the left in Figure 19.3 shows the decline of turnover of a large, post-Soviet company as the function of time: production declined and it soon became unprofitable, leading to the closure of the company – as happened with the Helylä ski, furniture and parquet company. After the bankruptcy (on the right in Figure 19.3), three new companies were founded and most utility and social service responsibilities were transferred to governmental bodies. Thus, public utilities and welfare provision were no longer responsibilities of any industrial company.

Figure 19.3 depicts the situation in which there are no possibilities to modernise production without starting from the very beginning. The reasons for this development track can be understood when the costs and revenues in the old and recently founded companies are compared (Figure 19.4). The crucial difference between traditional and recently founded companies is the way business is financially managed (Figure 19.4).

According to standard micro-economic theory TC=FC+VC, total cost is the sum of variable and fixed costs. These cost items can be used to describe the components of turnover, where TO=TC+P=FC+VC+P in which P equals profit. Total Cost represents the lowest aggregate expense required to produce each level of output q. Fixed Cost represents the total expense that is spent even when a zero output is produced. Fixed Cost usually includes contractual commitments for rental, maintenance, depreciation, interest, etc. Variable Cost represents all items of TC except for FC, such as raw materials, fuel, wages, transport, etc. In Figure 19.4, the cost components are described as a function of time in an outdated company in post-socialist country B, on the left of the figure. It is assumed that TO=TC=VC in the company in country B, which was the actual situation with many large Russian companies, founded during the Soviet era, in the late 1990s. On the right, the cost components of new, emerging companies are represented, and a cost structure such as this makes possible investing in and developing production.

The lower part of Figure 19.4 depicts the financial balance of local public services in the late 1990s. As observed in Russia, the reorganised social sector consists of expenditures on public services (EPS) which equals to the sum of the redistribution of federal taxes (RFT) and the subsidies paid by new companies (SPC) to local authorities.

Figure 19.4 clearly depicts the financial difficulties of a company within the Russian production context. In country B, the turnover of a company equals VC, which means that the company has no money for investments and R&D. This has been the case both in the Helylä company and Pechenga Nickel in the 1990s. Moreover, VC has not been very dependent on the pace of work. Wages have been paid, if possible, even though production has sharply declined. The total revenue has been distributed to workers. Pechenga Nickel is subject to the investment policy of Norilsk Nickel Inc. and dependent on its internal financial arrangements, but the described theoretical framework essentially holds true.

One crucial reason for this situation is that companies were formerly only production units, which did not themselves take care of allocating money for investments and R&D. This sort of management has continued in large Russian companies making them incompetitive in emerging market economy environments. A second reason originates from the recent investment climate in Russia. Investments have declined even faster than production, year after year from 1990 to 1998. Resource-based industries, as long-term and capital-intensive activities, have not succeeded in attracting even scanty investments. Both the structural and economic arguments became clear during the interviews with the managers of Pechenga Nickel in Zapolyarnyj as well as during the course of interviews in Helylä.

Public services are increasingly being outsourced from industrial companies and service provision tends to operate under the auspices of a local authority (Figure 19.4), though subsidised by companies in various degrees. In Figure 19.4, the description of financial arrangements of the newly founded public service sector is based on the interviews in Zapolyarnyj (Tatarinskii 1998) and Värtsilä (Purmonen 1998). In the latter case, a joint venture was established to produce sawn timber just near the Finnish-Russian border, some 50 km north of Helylä (Tykkyläinen 1998b). Insufficient tax revenues from federal authorities (RFT) have led local authorities to search for assistance from private sector companies (SPC) (Figure 19.4). Zapolyarnyj and Nikel in 1998 were still more like company towns in terms of public service provision.

Company strategies for economic recovering

Pechenga Nickel was still in operation in 1998 but was radically cutting expenditures and laying-off employees. It attempted to avoid drifting into the same situation as the Helylä company. The mining company has also successfully co-operated with the local authorities in transferring the responsibilities for public utilities to the local administration, especially as taking into account the scarce financial resources of local administration. The nickel company is still the major revenue source of the local administration (Tatarinskii 1998).

The company's labour force reduction programme will continue in 1999 in the same extent as in 1998. Fiscal rationalisation is also planned to be instrumented by outsourcing: some of the auxiliary departments of Pechenga Nickel will function as independent companies in the future. Regardless of even the most drastic cuts, the company is unable to put sufficient money aside for investments let alone for maintenance. For these reasons, external investors were deemed absolutely essential in order for management to avoid the unviable situation indicated by Figures 19.3 and 19.4.

Conclusions

Economic development in the Russian North is stifled by the former institutions from the Soviet era. This situation has led companies into a development trap: companies' revenues go to covering wages as well as social obligations and there is no appealing business environment for long-term investments. Without the radical and profound reform of companies' business environments, and especially financial management as referred to in this paper, all economic renewal attempts will take place within an economic quagmire, more often than not, spawning only bankruptcies and crises.

References

Bank of Finland (1998), 'Economic Indicators', *Russian Economy, The Month in Review* No. 7/1998.

Blatov, I. A. (1998), 'CEO of the Pechenga Nickel Company', Personal communication, 14.8.1998.

Kamkin, I. R. (1998), 'Executive director of mining, Pechenga Nickel Company', Personal communication, 14.8.1998.

Meert, H., Mistiaen, P. and Kesteloot, C. (1997), 'The Geography of Survival: Household Strategies in Urban Settings', *Tijdschrift voor Economische en Sociale Geografie* 88, pp. 169-181.

Mingione, E. (1991), *Fragmented Societies, A Sociology of Economic Life beyond the Market Paradigm,* Basil Blackwell, Oxford.

Neil, C., Tykkyläinen, M. and Bradbury, J. (eds.) (1992), *Coping with Closure: an International Comparison of Mine Town Experiences*, Routledge, London and New York.

Neil, C. & Tykkyläinen, M. (eds.) (1998a), *Local Economic Development, A geographical comparison of rural community restructuring,* United Nations University Press, Tokyo.

Neil, C. and Tykkyläinen, M. (1998b), 'Factors in local economic development', in C. Neil and M. Tykkyläinen (eds.), *Local Economic Development, A geographical comparison of rural community restructuring,* United Nations University Press, Tokyo, pp.309-317.

Nygren, L. and Karlsson, U. (1992), 'Closure of the Stekenjokk mine in north-west Sweden', in C. Neil, M. Tykkyläinen and J. Bradbury (eds.), *Coping with Closure: an International Comparison of Mine Town Experiences*, Routledge, London and New York, pp.99-118.

Purmonen, P. (1998), 'Director, Sirius Oy, Niirala', Personal communication 20.3.1998.

Tatarinskii, V. (1998), 'Deputy Director of the Pechenga District', personal communication, 14.8.1998.

Thrift, N. (1983), 'On the determination of social action in space and time', *Environment and Planning D* 1, pp.23-57.

Tsaplin, V. (1998), 'Technical manager, a former ski and furniture factory in Sortavala', Personal communication, 20.2.1998.

Tykkyläinen, M. (1998a), 'Theoretical and methodological underpinnings of the study of rural survival strategies in transitional countries', in M. Tykkyläinen, E. Varis, J. Oksa, M. Piipponen, I. Nágy, É. Kiss, and G. Mátray, 'Rural Survival Strategies in Transitional Countries', *University of Joensuu, Karelian Institute, Working Papers* 2/1998, pp.5-14.

Tykkyläinen, M. (1998c), 'A multicausal theory of local economic development', in C. Neil and M. Tykkyläinen, M. (eds.), *Local Economic Development, A geographical comparison of rural community restructuring*, United Nations University Press, Tokyo, pp.347-355.

Tykkyläinen, M. (1999), 'The emergence of capitalism and struggling against marginalization in the Russian North', in H. Jussila, R. Majoral and C. Mutambirwa (eds.), *Marginality in Space – Past, Present and Future*, Ashgate, Aldershot.

Tykkyläinen, M. and Jussila, H. (1998), 'Potentials for innovative restructuring of industry in Northwestern Russia', *Fennia* 176:1, pp.223-245.

Tykkyläinen, M. and Rautio, V. (1998), *Modernization of the mining communities in the Russian North: Options and local reactions*, A paper presented at the conference of Applications of Computers and Operation Research in the Minerals Industry, December 7-9, 1998, Kalgoorlie, Western Australia.

Tykkyläinen, M., Varis, E., Oksa, J., Piipponen, M., Nágy, I., Kiss, É. and Mátray, G. (1998), 'Rural Survival Strategies in Transitional Countries', *University of Joensuu, Karelian Institute, Working Papers* 2/1998.

World Bank (1997), *The State in a Changing World, World Development Report 1997*, Oxford University Press, New York.

Voronkov, Viktor (1995), Poverty in Modern Russia: Strategies of Survival and Strategies of Research, in K. Segbers and S. De Spiegeleire (eds.), *Post-Soviet Puzzles, Mapping the Political Economy of the Former Soviet Union, Vol IV, The Emancipation of Society as a Reaction to Systemic Change: Survival and Adaptation to New Rules and Ethnopolitical Conflict*, Nomos, Baden-Baden, pp.23-38.

Acknowledgements

The research pertains to the Research Programme of Russia and Eastern Europe of the Academy of Finland, projects No. 38812 and 60201. The financial support of the Academy of Finland is highly appreciated. Interviews in 1998 were undertaken during the Sortavala and Kola excursions, which were organised as part of Nordplus co-operation No. 0104/97 and the TMR-course contract ERBFMMACT960147 respectively.

20 Transition in the post-socialist countryside:
The restructuring of rural settlements in Hungary and Russian Karelia

EIRA VARIS

Introduction

Transition in different post-socialist countries varies both regionally and locally. Rural transition has a particular role to play in the process of post-socialist transition; the reorganisation of rural production and the household economy are of great importance to this post-socialist transition. Rural issues also have a significant role in the future development. They are essential for instance in the integration process of the European Union and the international development of utilising resources.

In this article rural transition is examined and compared in two post-socialist countries: Hungary and Russia. The article is based on local case studies made in two marginalized villages (Varis, 1998a). The villages of Hunya, located in the Great Hungarian Plain, Hungary, and Koivuselkä in the Karelian Republic, Russia. Both case study villages are small but have independent local governments. The populations of these two villages are approximately 900 and 400 inhabitants respectively. Economic activity is concentrated in agriculture in Hunya and forestry related activities in Koivuselkä. The fact that production in these two villages is dependent on natural resources has been decisive for local development in both villages and allows villages to be considered to be resource communities (cf. Neil and Tykkyläinen, 1998a, p.4-5).

Transition as a geographical concept

Theoretically the term 'transition' is applied to indicate the restructuring which takes place as the socialist system of society is being converted to the market economy. It indicates the change between two characteristically different social systems, an interlude between two qualitatively different social orders (Blom *et al.*, 1996, p.7). Transition can be considered a type of restructuring. It means that transition also includes the factors, which determine the restructuring. A specific element of transition is the movement towards a different social system.

Transition also has a spatial essence. It affects time, region and locality. According to the Stunde Null construct (cf. Paul, 1992; Tykyläinen, 1995, p.9-10) a fundamental change happens in a society - the collapse of socialism can justifiably be characterised as such a change - after which the structures of society start to renew themselves in a pronounced manner from new starting points. Concretely transition in Hungary is dated to have started in the system shift, which happened in 1989/1990; although it is impossible to define the exact time of the start as the origins of transition are the consequences of a chain of events. A similar type of transformation happened a little later in Russia when the Soviet Union ceased to exist at the end of 1991. The local consequences of transition are variable. When transition is examined through specific local features it is possible to gain an understanding of the importance of spatial differences and heterogeneity.

Transition is brought about by both the internal and external factors in localities (Figure 20.1). Therefore, in the explanation of local restructuring it is necessary to take into account the combination of different factors that interact with one another. General, sectoral and political factors are considered to be external factors of transition in this study (cf. Neil and Tykyläinen, 1998b, p.314), but the political measures are included in the analysis of the general and sectoral factors. However, the focus of this study is concentrated on local and human factors, as the research is designed as a locality study.

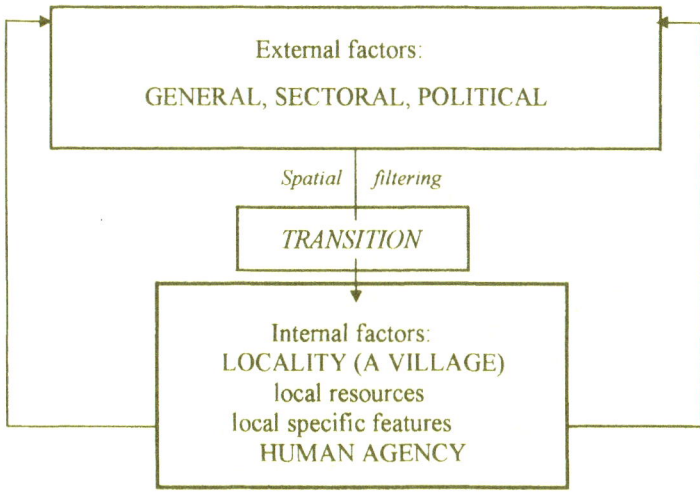

Figure 20.1   Factors of transition in a locality

External factors of rural transition

Collapse of the logic of socialism and the emerging market economy

The general factors are considered both international and national. The international factor, which contributed to the commencement of the transition process, was the collapse of the socialist regime and the repercussions of this collapse, such as the disintegration of the socialist trade block and the disruption of socialist ideology. This international factor represents a change in the political economy at the global level and thus concerns both of the post-socialist countries under examination.

Transition at the national level can be estimated by evaluating the changes in the structures of the political economies. On the national level the political economy of Hungary is undergoing a transformation from plan ideology to market rationality (cf. Appelbaum and Henderson, 1992, p.18-21). In other words, the old socialist ideological dogma, incorporating state ownership and control is being replaced by the regulatory function of the state with smoothly functioning markets (*ibid.*). The manifestations of this process are, for example, the privatisation of state-owned property and, at the ideological level, the abandonment of the one-party monopoly and the rise of democracy.

In Russia the foundations of the political economy are changing from a plan ideology to a market ideology. Within this market ideology political decisions are taken in the context of the free market (*ibid.*). In order to function properly market ideology needs to present possibilities for private ownership and the rise of democracy. However, structures still exist in Russia, which originated from the old plan ideology, for instance state land ownership or even the state farms in agriculture or the fishing kolkhozes. Thus, features of plan rationality also exist. The distinct transformation from one system to another has not happened. A part of the old system has been radically changed, but on the other hand part of the old structures have not (yet) been replaced.

During the phase of transition all transitional countries have faced a lot of economic, political and social difficulties and problems (Table 20.1). In both the countries under consideration in this study transition has brought considerable problems to the national economy. Because of the experience of hyperinflation one could be even talk about the existence of an economic crisis in Russia. Transition has also caused social problems that did not exist in socialism. For instance, unemployment is a new occurrence that did not exist during socialism in either Hungary or Russia.

Table 20.1 Factors of transition in the case study villages at the general and sectoral levels*

Level	Hunya, Hungary	Koivuselkä, Russia
general	– 'system shift' 1989/1990 – democratisation – privatisation • economic problems • high inflation • unemployment • adapting to the EU	– the Soviet Union collapses 1991/1992 – democratisation – privatisation • economic crisis • hyperinflation • unemployment
sectoral	– break up of socialist co-operatives – compensation and redistribution of land • newly-formed co-operatives • re-establishment of peasantry	– privatisation of the forestry complex • establishment of joint-stock companies • break up of the state social sphere

* – event, • result

Changing industries

Sectoral factors vary from one industry to another depending on the political measures, the functioning of markets and the development of technology (see Table 20.1). The aims of economic effectiveness rule the development of production sectors. During socialism the aims were implemented by political planning and decision-making. In the transition period changes are increasingly dependent on market forces. From the viewpoint of a resource community this means, that transition is the search for new opportunities in resource-based production. In addition, technological progress like the introduction of modern equipment affects restructuring.

In the Hungarian case, the industry concerned is agriculture, as the case study has been carried out in an agricultural village. Hungarian agriculture, owing to favourable geographical conditions and historically accumulated know-how, has played an important role in the national economy, regional division of labour and foreign trade. On the other hand, it has also been a local form of organisational structure in rural areas and thus affected the regional structure.

During socialism the production was organised with state farms and co-operatives in Hungary. An agricultural co-operative operated in the case study village of Hunya. It determined the community's position in the regional division of labour. Therefore the development of the community has been reflected by the development of the production sector, i.e. agriculture. After the system shift the socialist co-operatives were broken up. Collectivised land was compensated and the land was redistributed. However, the majority of these co-operatives were later re-established. Private peasantry again became both conceivable and achievable after the socialist period and a new process of peasantisation commenced. In the near future agriculture as a production sector will have a significant meaning for Hungary's integration into the European development framework because of its intentions to join the EU (cf. Varga, 1998, p.118-119).

With regard to the case study village of Koivuselkä the production focus has been the forestry sector in the Soviet Union and later, in Russia. Forestry has been the dominant industry in Russian Karelia because the forest is its main natural resource. In the regional division of labour of the Soviet Union the role of Russian Karelia was to produce and process wood products. During the soviet regime the forestry sector operated as a huge production complex based on natural resources (cf. de Souza, 1989). Rural settlements functioned as a part of this. Thus, the development of the forestry sector has also played a role in the formulation of the regional structure. Koivuselkä was a part of this complex by through its *lesopunkt* production unit, i.e. a logging unit.

After the collapse of socialism the forestry complex was partitioned. The components of it were privatised and joint-stock companies were formed out of them. People were able to take part in the privatisation through a voucher system. In spite of this the ownership of the new companies by members of the local community remained low. Scattering the forest complex also meant the break up of the state social sphere. The state organs of the forestry companies had previously taken care of the social services in industrial settlements. However, new private companies no longer have this responsibility. This means that social welfare decreased dramatically and the standard of living declined. Furthermore, during transition changes at the international and national level concerning the forestry sector, including implementation policy, also has strong local effects (for instance Oksa *et al.*, 1997).

Internal factors of transition

Transition of the local communities

The effects of external forces are filtered down to the local communities through the local conditions and are further shaped by the local actors (cf. Tykkyläinen, 1998, p.350). Therefore the internal forces, such as local conditions and local actors, also have to be also taken into consideration when analysing the restructuring at the local level (cf. Varis, 1998a). The responses to the pressure of restructuring are dependent on specific local features like labour skills, communications, the industrial mix or the opportunities which exist to exploit resources (Tykkyläinen, 1998, p.348). They may also be dependent on local initiative, cultural features, mentality and demographic structure. The responses of the local level are contingent upon the potential available in that place. In this study the local level is analysed from two standpoints. On the one hand the resource community is examined as a locality and, on the other, the analysis highlights the local consequences of restructuring by studying the human factor which is represented here by households (see Figure 20.1).

Privatisation with its consequences has been one of the most important events, which has affected the institutional transformation in rural communities both in Hungary and Russia (Table 20.2). The socialist co-operative

Table 20.2 Factors of transition in the case study villages at the local and individual levels

locality	• socialist co-operatives are broken up and newly-formed co-operative are established • land auctions • land ownership move into private possession • new group of agricultural entrepreneurs • new smallholders • cultural services dissipate • commercial services become private • local government is strengthened • new people move into the village	• loss of position in the forestry production complex • reduction in employment and decision-making possibilities • deterioration and reduction in service provision • high cost of consumer goods • belief in the future of the village weakens • new people move into the village
individual	• liberalisation of mental atmosphere • political freedom • notion of land-ownership becomes possible • individual initiative and entrepreneurship become possible • economic problems • for some households their secondary economy activities become their primary economic activities, for other households secondary economic activities remain important • social conflicts arise as a consequence of the new situation • uncertain future	• liberalisation of mental atmosphere • political freedom • unemployment • economic problems • the importance of self-sustainability increases • social problems increase • social conflicts as a consequence of the new situation become evident • uncertain future

of Hunya was dismantled, but after a short while a new co-operative was established. Hunya suffered from this political decision because the village's socialist co-operative functioned well. Thus, real property, such as building, machinery and equipment, was wasted. Fortunately for the local community the newly formed co-operative took on some of the social functions of the old co-operative. It was the wish of the management of the co-operative. The most important of these social functions were the employment of local people and the maintenance of living standards for pensioners.

The system shift resulted in a reorganisation of work. The newly formed agricultural co-operative became the successor of the socialist co-operative and was, therefore, the most important employer in the village. The sewing co-operative, which had also existed during socialism, provided employment for women.

Transition also made private entrepreneurship possible and some new enterprises were established in the village of Hunya. These enterprises are primarily involved in agriculture and services that support this economic activity. Some commercial services were also established. The reorganisation of the local government structure has transferred more decision making power to the local level. Furthermore, a multi-party system became possible. These changes made people feel that their own potential for decision-making grew as a consequence of transition.

The peripheral forest village of Koivuselkä lost its position in the hierarchical organisation of the new forest company after the privatisation of the forestry complex. The logging unit *lesopunkt* was closed and in its place a less important forestry unit type, *masterskij utshastok*, was established. This reorganisation of the forestry activities meant that some operations in Koivuselkä ceased. For instance, the local office of the forestry company was closed. Thus, some employment loss occurred. As a result of this rationalisation taking place in the organisation of the forestry company some of the decision-making concerning forestry activities in Koivuselkä was also lost. This gave local inhabitants even fewer opportunities to influence the processes that affected them.

The significant changes also took place in the social sphere of the village of Koivuselkä. The forestry company no longer took responsibility for the shop, the canteen and the farm. The shop continued to function under the public administrative organ. Some private joint-stock companies took on the responsibility for some of these services on a small scale. Private enterprises were not established in the village.

New actors and local survival

The changes which transition has brought have encouraged the emergence of new local actors in both the case study villages.

Several new actors in Hunya

The agricultural co-operative had represented the major socialist institutional source of employment in Hunya. The agricultural activities undertaken within the framework of this co-operative formed the primary industry for the village. Villagers were either members of the co-operative or farm workers in the co-operative. As a consequence of privatisation this situation changed. Those who had been members of the co-operative received a portion of the co-operative property during the privatisation process. Some of the villagers also received a share of the co-operative as restitution. The change in land and property ownership created new types of institutional forms in the agricultural sector. These three groups can also be considered as new actors who have emerged as a result of transition:

1. The newly formed co-operative: people are members and work for wages. They have leased their land to the co-operative (co-operatives are not allowed to own land).
2. Private agricultural enterprises: people work for wages in an enterprise that is usually managed by the owner.
3. New family farms: the landowners are farmers.

In 1996 the newly formed co-operative consisted of 56 households with 82 working members and about 180 pensioner members. In 1989 the previous co-operative had had over 212 active members and 275 pensioner members (according to data from the co-operative archive). Thus the transition process reduced the number of members. This form of action, which can be described as the strategy of continuity, played a significant role among the village households. Work for an employer other than the co-operative was also very common. In the strategy of paid work the work itself was not usually agrarian. This strategy was mainly relevant to women. One important actor in the provision of employment for women in Hunya both during socialism and transition was a sewing factory. This factory has been operating since the beginning of the 1980s, and in the period when this research was undertaken it was a KFT-type of enterprise (the closest English equivalent a limited liability company, see Varis, 1998b, p.297) and offered work to about 25 women. The owners of the sewing factory themselves came from outside the village. Additionally village services such as the school, shops and offices primarily provided employment for women.

Those active-age households, which did not stay in the co-operative but, still gained their livelihood from agriculture could either establish a family farm or a private enterprise; thus they adopted the strategy of exploiting the new opportunities. However, this group was rather small, contrary to what was estimated and anticipated in the state politics (cf. Hann, 1996, p.39). An agricultural company, a family co-operative and some private agricultural enterprises and lorry firms were established. The registered enterprises had few owners and employed workers. These enterprises were mainly involved in cultivation; they also supplied some machinery services. They can be characterised as service enterprises. There were only a few 'pure' smallholders who were occupied solely with farming their land.

The most common strategy of the households was the mixture strategy. Generally, this composition of livelihood consisted of paid work in the cooperative whilst working for another employer, usually in the sewing factory or public services. The households which performed this combination of employment did not change their primary economic activity from that which they were involved during the time of socialism. Furthermore, there were some 'mixed' households, where one family member stayed in the cooperative and another established a private enterprise or a family farm. These households were cautious, keeping 'one foot in the door' of the cooperative whilst at the same time testing the new opportunities offered by transition through the processes of private entrepreneurship. If the household had more than two active-aged members it was also possible to adopt all available strategies to maintain the primary economic activity.

Social networks as the most important actors in Koivuselkä

The participation of new actors in the Koivuselkä transition has in the role that the working unit; *masterskij utshastok* plays being reduced. Instead the village soviet had to take more responsibility for the survival of the village. The role of social networks has increased. Thus these different types of social networks may be regarded as the most important new actors (cf. Oksa, 1998, p.36).

Because of the drastic changes of the institutional providers in the village the survival strategies of the villagers have been founded on the notion of self-sufficiency. As a result of the huge economic problems and the confused social situation the society is no longer in a position to support the livelihoods of the villagers. Thus the cultivation of private household garden plots and the gathering of natural resources became increasingly vital to the local people. Many households are almost self-sufficient in their household economy and are living in a situation of autarchy. Small plots of land in the countryside are now also more appreciated by urban dwellers.

The significance of the social networks has increased (cf. Rose, 1994). The connections to rural areas enable these people to take advantage of the available natural resources. Picking berries and mushrooms, gathering of herbs and fishing are vital for inhabitants of Koivuselkä and their relatives (Table 20.3).

Table 20.3   Old and new actors and survival strategies in Hunya and Koivuselkä

	Hunya	Koivuselkä
Old actors	♦ agricultural co-operative ♦ sewing co-operative ♦ village soviet	♦ logging unit of lesopunkt ♦ village soviet
New actors	♦ new agricultural co-operative ♦ sewing co-operative ♦ private agricultural enterprises ♦ family farms ♦ local government	♦ sub-unit of lesopunkt ♦ village soviet ♦ social networks
Survival strategies	♦ continuity ♦ paid work ♦ exploiting new opportunities ♦ mixed strategy	♦ self-sufficiency

Concurrent with the first, official economy a second, unofficial economy was in operation in both case study villages during socialism. This second economy relied on the individual activity and work of the local people. The second economy in Hunya specialised in the farming of maize seeds in privately controlled plots. Today many farmers gain the majority of their production from these plots. The only difference is that the former secondary economy has now become the source of primary economic activity for the household. In Koivuselkä the products of the secondary economy came from farming and forestry. Thus this secondary economy changed its role to become the primary economy.

Comparing transition in rural Hungary and Russian Karelia

The biggest institutional difference between Hungary and Russia is in land ownership. Private landownership has returned in Hungary. As a consequence of this a structure of private entrepreneurs and peasants has developed. In addition, the leasing of land has become a significant additional source of income, particularly for pensioners in the countryside. The privatisation of land has also had an influence on the mental atmosphere and has given a new beginning to the concept of private entrepreneurship.

In comparison, in Russia land is still owned by the state. The Russian Parliamaent (Duma) has confirmed the Land Act according to whom it is not legal to buy or sell land now or in the future (Kozlov, 1997). At present the Act does not have the confirmation of the President. However, should it be confirmed private land ownership will not even emerge in the future.

Furthermore, in Russia, on contrast to Hungary, there is no longer anyone who would remember the period in which private land ownership was possible. Moreover, the Russian Karelia has never really had private land ownership in the modern sense. Thus, the rights concerning ownership are substantially different in these two areas.

Transition has brought new situations that did not exist during socialism. The most significant to affect the populations of these areas is unemployment that has, for the most part, spread very rapidly. The unemployment rates vary regionally, but the common feature for the transitional countries is the fact that rural areas have been badly hit by this phenomenon (Kovách, 1997). In terms of unemployment the peripheral village of Hunya is an exception. The unemployment rate of this village is only two per cent. The reason behind this lies in the history of the village. Before the collectivisation there were hardly any landless workers in the village. Thus, after the system shift the majority of the villagers got land compensation and the possibility either to start a family farm or to join an agricultural enterprise or to join the new co-operative. In contrast, Koivuselkä has faced high unemployment. The estimated rate according to household surveys undertaken in the village is that the level is between 20 and 30 per cent (Klementev et al., 1996, p.201). This level of unemployment may be associated with the rationalisation of the forest company after privatisation.

Economic problems, particularly rapid inflation, are typical problems in the transitional countries. This is also reflected in people's everyday lives. Individual savings have fallen or, in many cases, totally ceased. The situation is such that people now have to compromise when purchasing goods and services because prices have increased at a far greater rate than either wages or pensions. In Russia problems have been encountered by those receiving wages and pension, this situation is present in the village of Koivuselkä. The claims may be delayed for months and this has forced people to switch to a lifestyle that relies on greater self-sufficiency.

Changes in the local social sphere, for the most part, had an impact on the local people through the changes that have taken places as regards work patterns. During socialism socialist enterprises also supported the physical and mental culture of a community. It is no longer the duty of the enterprises to support these aspects of the community. Thus, for the local inhabitants it seems as though mental capital is being lost. Villagers have tried on a voluntary basis to organise some hobbies club for the children at least. However, some services, such as shops, have improved because new shops have appeared creating competition.

The village soviet and the district administration maintain the remaining public services in Koivuselkä. The cultural services have also declined in Koivuselkä, this is considered to be a drawback amongst the villagers. The situation has changed in form since the first years of transition when there was a lack of consumer goods. Now, however, food and goods are quite expensive. Poverty has increased and money is only spent on necessities.

Countryside has offered an alternative living place for many people who have been faced with the consequences of transition in the towns. In spite of their remoteness both Hunya and Koivuselkä have new inhabitants who have migrated from the towns. Some retired families and families with children who were made redundant in the towns have moved to Hunya. The modern facilities and the cost of houses in Hunya have encouraged families to sell their residences in the town and migrate to the countryside.

In comparison, the flats of Koivuselkä are in such a bad condition that they are not attractive propositions, as such they cannot be considered as a particular incentive to those who choose to move to the village. Those who have moved to the village are predominantly families with several children that have been affected by the high level of unemployment in the towns. The motives for their migration include the existence of a self-sufficient economy in the village. Even though the number of immigrants to both villages is not large they have increased the number of school children in the villages' schools and, in this manner, have improved the vitality of the villages.

Both Hunya and Koivuselkä have faced similar events associated with the transition process such as privatisation, economic problems and unemployment. The different local conditions have played a role in determining the different possibilities that face the local communities. Thus the consequences of transition in these two villages have varied. Despite the transitional difficulties a new kind of local activity has emerged in Hunya. The village is ready to face the challenge of its future development. In contrast, the relationship between the regional division of labour and Koivuselkä is weakening. Unless there are dramatic changes in the village, an economy based on self-sufficiency will be the only strategy available to the village people for survival.

Conclusions

The rural areas of Eastern Europe faced both great change and turmoil during socialism. Today they are again undergoing transformation with the transition from socialism to the new social order and movements towards a more market based economy. Continuous changes have created a situation in which the local people must create their own survival strategies in order to adapt to the changes.

One can distinguish various development paths in different localities. The specific local features, like a willingness to exploit the local resources in a new manner, have proved to be important for survival. Endeavours in market-based entrepreneurship in Hungary have proved to be strong. On the other hand, in Russia the rural households are returning to a self-sufficient autarchy in order to cope with transition. A strong effort to westernise and to be separated from the former socialist block is exists in Hungary. Hungary is, for example, very eager to participate in the widening process of the European Union. Developments towards such integration will bring new challenges, particularly for Hungarian agriculture and thus for the rural areas. The social and economic problems which transition has brought have had a destabilising effect on Russia and as a consequence the people have experienced mass shortages, particularly in the peripheral areas.

In these transitional conditions the patience of the local people and their ability to adapt will play a vital role in maintaining the social order. Adapting to the changes, which have had mainly a negative effect on the lifestyles of these people, will involve a strong will on the part of the people if the transition process is to succeed. Changing the outlook of the rural people according to the new conditions takes time. The decisive question will be whether or not the achievement of democracy will be sufficient to maintain the patience of the people.

References

Appelbaum, R.P. and Henderson, J. (1992), *States and Development in the Asian Pacific Rim*, Sage Publications, London, p.320.

Blom, R., Melin, H. and Nikula, J. (eds.) (1996), *Between plan and market: Social change in the Baltic States and Russia*, Societies in transition 6, Walter de Gruyter, Berlin, p.182.

Klementev, Je., Oksa, J., Polevshikova, N., Rannikko, P. and Varis, E. (1996), 'Local impacts of restructuring - A case study of a forestry settlement', in E. Varis, and S. Porter (eds.) *Karelia and St. Petersburg. From Lakeland Interior to European Metropolis*, Joensuu University Press, Joensuu, pp.191-206.

Kozlov, A. (1997), *Sobstvennost na lesa v uslovijah stanovlenija rynotshnyh otnoshenii*, paper presented in the seminar 'Taiga Model Forest and social sustainability', in Joensuu December 1997.

Neil, C. and Tykkyläinen, M. (1998a), 'An introduction to research into socio-economic restructuring in resource communities' in C. Neil, and M. Tykkyläinen (eds.), *Local Economic Development: A Geographical Comparison of Rural Community Restructuring*, United Nations University Press, Tokyo, pp.3-26.

Neil, C. and Tykkyläinen, M. (1998b), 'Factors in local economic development', in C. Neil, and M. Tykkyläinen (eds.), *Local Economic Development: A Geographical Comparison of Rural Community Restructuring*, United Nations University Press, Tokyo, pp.309-317.

Oksa, J. (1998), 'Conceptual framework of a village', in M. Tykkyläinen, E. Varis, J. Oksa, M. Piipponen, I. Nagy, É. Kiss, and Gy Mátray, *Rural survival strategies in transitional countries. An introduction to the comparative study of localities in Northwestern Russia and Hungary*, University of Joensuu, Karelian Institute, Working Papers N:o 2/1998, pp. 28-39.

Oksa, J., Rannikko, P. and Varis, E. (1997), 'Sozio-ökonomische Bedeutung der Forstwirtschaft in Russisch Karelien', *AFZ Der Wald*, 52:14, pp.747-748.

Paul, L. (1992), 'Developments in Rural Areas in Eastern Europe Since the Second World War: a overview', in P. Huigen, L. Paul and K. Volkers (eds.), *The Changing Function and Position of Rural Areas in Europe*, Nederlandse Geografische Studies 153, Faculteit Ruimteljike Wetenschappen Rijksuniversiteit Utrecht, Utrecht, pp.101-108.

Rose, R. (1994), 'Getting by without government: Everyday life in Russia', *Daedalus* 123:3, pp.41-62.

Souza de, P. (1989), *Territorial production complexes in the Soviet Union - with a special focus on Siberia*, Department of Geography, University of Gothenburg, Serie B nr 80, Gothenburg, p.257.

Tykkyläinen, M. (1995), ''Stunde Null' and Pocket Theory of Development in the Socioeconomic Transition in Russia' in M. Tykkyläinen (ed.), *Russian Karelia – An Opportunity for the West*, pp. 9-23. University of Joensuu, Human Geography and Planning, Occasional papers 29, Joensuu.

Tykkyläinen, M. (1998), 'Multicausal theory of local economic development', in C. Neil and M. Tykkyläinen (eds.), *Comparing Local Economic Development. A Geographical comparison of rural community restructuring*, United Nations University Press, Tokyo.

Varga, Gy. (1998), 'Hungarian Agriculture and the EU', *The Hungarian Quartely* 39:150, pp.108-119.

Varis, E. (1998a), *Syrjäkylien murros Venäjän Karjalassa ja Unkarissa* (*Transition of remote villages in Russian Karelia and Hungary, A study of restructuring of post-socialist countryside and survival of resource communities*), University of Joensuu, Publications in Social Sciences Nr. 33, Joensuun yliopisto, Joensuu, p.312.

Varis, E. (1998b), 'Facing transition in rural Hungary - a case study of an agricultural village', *Fennia* 176:2, pp.259-300.

Acknowledgements

The project pertains to the Research Programme on Russia and Eastern Europe of the Academy of Finland, project No. 38812. The financial support of the Academy of Finland is highly appreciated.

21 Residential and settlement issues of forest sector communities in the Republic of Karelia, Russia

MINNA PIIPPONEN

Introduction

This article discusses the contemporary changes taking place in local communities involved in the forest sector of the economy in the Republic of Karelia, Russia. It presents underpinnings of the study plan on the subject. The focus of the study is to address the following questions: how the reactions and actions of local and supralocal actors modify the development of residential and settlement issues during transition? And, what kind of influence do these developments have on local community and settlement formation? The article then provides a closer examination of two case study villages in response to some preliminary considerations regarding the starting points for the processes of change.

The economy of the Karelian Republic is founded on the utilisation of forest resources. Since the early years of the institutional development of the former Soviet Karelia in the 1920s the use of forest resources became an important element of the Soviet Union's efforts to industrialise this area quickly (Autio, 1997). The role that the area played in the production system was carried out according to the common principles of the planned economy, this was facilitated by the building of a production complex based on the natural resources of the area (territorial production complexes see Souza de 1989). This development culminated in the mid 1980s, when organisations and state enterprises involved in forestry and forest industries became part of one large production complex called Karellesprom, an abbreviation of 'Karelian Forest Production Complex' (Oksa and Saastamoinen, 1995, p.99). Today the forest sector, timber harvesting, wood processing and chemical forest industry, accounts for more than 40% of the total industrial production and 48% of the industrial employment in the Republic of Karelia (Respublika Karelija v tsifrah, 1997, p.22).

The use of dispersed forest resources has influenced the settlement system of the area. In addition to one large town there are 12 smaller manufacturing towns, and many of the local economies of these smaller towns are specialised in forest related industries (Oksa and Varis, 1994, p.61-62). In the beginning of the post-war period a great deal of settlement construction occurred in the rural areas as a result of the development of the forest sector. In 1957 324 new forestry settlements specialised mainly on wood harvesting were built (Klementev and Kozhanov, 1988, p.17).

This development can be said to have generated so called 'resource communities' in which the economic base is gained from and a living is made from specialising in the utilisation of natural resources (Tykkyläinen and Neil, 1995, p.32). The local community tends to function around one production unit or enterprise. The powerful influence of forest enterprises on the lives of whole communities has been a common feature in Russian Karelia. The former state enterprises did not only provide the work place for the residents; they also had social responsibilities. The enterprises provided the infrastructure, social and cultural services and housing.

Compared to the previous development, the transition period after the collapse of the Soviet Union has altered the foundations of village life in many communities. The ability of the forest sector to modernise and reorganise its production is an important part of the modernisation of the whole economy in the contemporary Karelian Republic. The former state enterprises have been reorganised, frequently into private joint-stock companies. However, new stock companies are no longer able and willing to carry the costs of the social responsibilities, which used to be a significant part of their budgets (Oksa and Saastamoinen, 1995, p.99-101). As a result, a major redistribution of social responsibilities in local communities is currently taking place.

Act of residence and settlement issues - framework for the study

This study aims to provide a clearer understanding of the spatial behaviour and decision making of principal actors regarding residence and its impact on settlement formation. Processes of change and their consequences are studied at the local level. Thus, the study has its methodological basis in locality studies (see Kortelainen, 1996, p.24-25 and Rannikko, 1989, p.22). In this case, three different communities involved in the forest sector of the economy in the Karelian Republic will be studied. Two of them are settlements in rural areas based on forestry and the third is a small manufacturing town in which the economic base is dominated by the pulp industry.

What, then, are the important elements for the study of residence and settlement issues, and how to connect these two concepts within this study? The thoughts of Kemeny (1992, p.78) offer a suitable starting point for these considerations. He has argued that with regard to housing, instead of talking solely about dwellings or households, the wider social implications of housing should be taken into consideration. He divides housing into three dimensions of increasing scope of social structure. They are household, dwelling and locality - these, he argues, form the phenomenon of residence, which is also regarded as a preferred term instead of housing.

This is supported by the considerations of Carter and Jones (1989, p.38-44). They discussed the externalities of residential location in their definition of three levels of satisfaction which housing is expected to confer upon its occupants. They stress that any assessment of residential satisfaction would be incomplete if it considered only the dwelling house in isolation from the surrounding environment. The place of residence is conceived as a location, which exerts a profound influence on the well-being and life changes of its occupants.

Studies of mining communities in Western Australia have illustrated that settlement formation may be conceptualised with the use of two main factors (see Tykkyläinen, 1994, p.202-203). From the residents' point of view, settlements as communities are places where it is possible to satisfy the needs of everyday life. They contain the sources of livelihood and services, as well as the social and cultural aspects of life. People prefer to choose a mode of residence that satisfies their needs best. The formation of a settlement thus depends on individual decision making about how to satisfy different needs. The other main factor is the economic actor, i.e. the enterprise that formulates its production processes and economic activities. In the illustration provided by Tykkyläinen individual preferences and needs are emphasised as prerequisites for finding settlement systems which are both adaptable and acceptable. However, physical conditions, societal well-being, societal values and resources and the pre-existing infrastructure provide the backdrop and preconditions for these decision-making processes. Thus, different societies and the development phases that these societies have reached offer a variety of outcomes as a result of individual decision making.

According to Tykkyläinen (1998, p.8-9), both individuals and communities have a role to play in the restructuring and transition processes. This creates the conditions for an environment of interaction and the development of causal relationships between the supralocal transition processes, for example between processes of privatisation and movements towards a freer market and the local development processes. The influence of these interactions on the development of a community is considered in the multicausal theory of local development (Tykkyläinen, 1999). In this theory the supralocal development processes are identified as general, political and sectoral (sector of industry) factors. Their interplay generates changes in local communities, but the results vary according to place because of the specific local factors. The local factors are considered to be a bundle of locally derived processes and agencies primarily determined by human actions. But they are also influenced by the attributes of the physical environment and local resources. Local factors are based on unique configurations of population, infrastructure, local economy, institutions and culture. The human agency is considered central in reacting to processes of change. It is constituted both of individual human behaviour and the behaviour of various coalitions of individuals, which have the aim of promoting development.

The multicausal theory offers the framework for the consideration of changes in residence and settlement formation during transition. The general, political and sector specific supralocal transition processes generate changes which are crucial for decision making relating to residence. For example, they result in changes in property rights and forms of tenure, and, in the modernisation of enterprise structures, production activities and formation of markets. The choices of individuals and households concerning their residences are ways to react to the processes of change in supralocal and local factors. As Tykkyläinen and Neil (1995, p.33) have illustrated, restructuring is not just an economic response to economic impulses, it is also linked with institutional, behavioural and socio-demographic processes.

In this study emphasis will be placed on the reactions of residents. Ties between the local community and its residents are important. These allow consideration to be given to how the contemporary communities serve the everyday needs of their residents. From this it may be possible to understand just how willing and how capable residents are of adapting to the processes of change. Of central importance is also that consideration is given to the reactions of the institutions, the forest enterprises and the local authorities. An understanding of the responses of these actors will help us to understand their interests and motives with regard to settlement and residence issues. It should also be possible to determine how capable they are of meeting the demands of change and what obstacles provide a barrier to their adaptive capacities. The reactions will be considered in the context of the preconditions and opportunities available in the Russian North.

The Forestry settlements of Koivuselkä and Matrosy

To begin giving attention to the processes of change consideration of the practices and structures that the Soviet society created regarding residential and settlement issues in local communities is needed. The preliminary considerations which are presented in the next chapters about residence and settlement issues in two of the intended case villages are based on empirical material which has been collected on several occasions between 1994 and 1998 in connection with different research projects regarding rural development in transitional countries and social sustainability.[1]

The settlement of Matrosy is located 35 kilometres west of the capital of the Karelian Republic, Petrozavodsk, by the St. Petersburg - Murmansk highway. The more remote settlement of Koivuselkä is located about 40 kilometres to the North East of the shores of Lake Ladoga, but by road it is about 150 kilometres from the capital. The populations of these two villages are approximately 950 and 380 residents respectively.

In general, a distinctive principle of settlement planning in the former Soviet Union was the desire for more 'town-like' settlements. Such settlements were considered to be more efficient and suitable for the industrialised socialist state even in the rural areas. (Pallot, 1987, p.329-330, p.338-343.) In the former Soviet Karelia the main role of the settlement was to provide the labour force for the forest sector, this was particularly the case for forestry in the more rural areas. Autio (1997, p.140-147), in her study about the development of forestry on the area in 1926 - 1932, refers to the idea that forest work settlements were established to act as a kind of industrial centre for the rural areas and could be used as a means of proletarianising the labour force. More practical reasons for establishing such centres included the need to improve the availability of labour and the productivity of the forest sector in order to meet the increasing demands of the planned production quotas set by the central government. In spite of the efforts to improve the productivity of the labour force, unhealthy and inferior living conditions were recorded in many places during the first years of the development of the forest sector.

With regard to the development of the settlements wood harvesting units, lesopunkty, of forest enterprises, lespromkhozy, have been decisive in the villages of Koivuselkä and Matrosy. The units gradually became part of the same Shuja-Vidanskij lespromkhoz, known today as the privatised Shujales-company. As forestry settlements, the villages are products of different time periods. In the case of Matrosy, the settlement was founded in the old village of Matrosy. Originally the economic activities in the village were made up of land cultivation and animal husbandry. However, during the first decades of the Soviet era the village began to grow and develop as a forestry settlement. Forestry had in fact flourished in Matrosy prior to World War II. During the construction boom of the forestry settlements in the early post war period a totally new settlement named Koivuselkä was established in 1949. The new settlement was located in a previously unsettled area in the neighbourhood of the old Karelian agricultural villages, which formed part of the village soviet of Pulchejla in the Soviet administrative structure.

The industrialisation process and the increasing role of the forest sector generated a flow of immigrants to the former Soviet Karelia. The shortage in the labour supply for the growing wood harvesting industry was already creating problems in the late 1920s. Solutions addressed to reduce this problem included the recruitment of workers from other parts of the Soviet Union, recruitment of convicts from labour camps, the collectivisation of agriculture and an obligation placed on new collectives to take part in forest work. In order to improve the techniques and productivity of the workpeople from Finland, as well as Finnish emigrants in North America, were induced to move to Soviet Karelia and work in the forests. (Autio, 1997, p.126-137, p.149.)

Table 21.1 illustrates how the development of the forestry has influenced the structure of the population in the case of Koivuselkä. Koivuselkä also drew labour and new residents from the surrounding Karelian villages, which disappeared gradually after the agricultural kolkhozes were closed in them in the 1950s (Klementev et al., 1996, p.193-195). In the case of Matrosy, immigrants from Finland and Finnish emigrants in North America contributed to the labour supply. This village acted even as one of the model type of units which were founded in order to test and spread the most productive techniques to other units in the early years of the development of the forest sector (Autio, 1997, p.154).

Timber harvesting was at its highest in the former Soviet Karelia in the 1960s when approximately 20 million cubic metres were harvested annually. Since then the volume has decreased. In 1993 the annual harvesting volume was about 6-7 million cubic metres in the Republic of Karelia. (Myllynen and Saastamoinen, 1995, p.45.) During the peak of the 1960s the predecessor of the present Shujales-company harvested annually 700 000 cubic metres of round timber from its different wood harvesting units (Klementev *et al.*, 1996, p.196). In 1980 the total volume harvested had fallen to approximately half the level seen in the 1960s peak (Pladov, 1998). This situation of decline was also apparent in the units of Koivuselkä and Matrosy from the end of the 1960s onwards. The nearby forests became exhausted or transportation and cutting technologies were further developed reducing the overall demand for labour in these logging units, and generating out-migration from the settlements (see Klementev *et al.*, 1996, p.196).

Table 21.1   Population of Pulchejla village soviet by nationality, (%)

	1926	1959	1970	1993
Karelians	98,6	54,3	54,8	61,0
Finns	0,3	10,4	8,3	19,9
Russians	1,1	14,6	24,1	11,9
Belorussians	-	16,4	10,5	4,1
Others	-	4,3	2,3	3,1
Total	100	100	100	100
Total population	783	1532	964	387

Source: Klementev *et al.*, 1996, p. 198.

The decline in logging operations has continued into the 1990s. In 1990 the production of Shujales-company was approximately 300,000 cubic metres, by 1997 this had fallen to 200 000 cubic metres (Pladov, 1998). This decline in production has changed the status of the production units in the settlements. The latest change was in 1995, when the lesopunkt of Koivuselkä was demoted to act as a subunit, masterskij uchastok, of the lesopunkt in Sodder. In the case of Matrosy such a change had already taken place, it is now a sub-unit of the lesopunkt in the settlement of Chalna.

Physical living conditions

After the logging operations had been established in the settlements of Koivuselkä and Matrosy the enterprise started to provide accommodation for its labour force. Today, the physical living conditions in this type of accommodation are a common subject of complaint amongst the residents of both villages (Tables 21.2 and 21.3).

Table 21.2   Satisfaction with living conditions in Koivuselkä

Tenure form	Satisfied (%)	Not satisfied (%)	Number of answers
Private detached / semi-detached houses	84	16	25
Detached / semi-detached company houses	45	55	40
Company apartments and barracks	38	62	32

Source: Household interviews in July 1994.

Table 21.3   Interviewees' opinions about the most important problems in Matrosy

Classification of the mentioned problems	Number of answers
Work	12
Living conditions	12
Services	27
Local authorities	2
Food	3
Environment	3
Can not say	3
Total	62

Source: Household interviews in October 1997.

In relation to housing provision, one of the main problems dates back to the construction and maintenance of wooden houses mainly for temporary use by the enterprise. Detailed instructions became to regulate the construction of the forestry settlements (Tehnicheskie ukazanija..., 1964, p. 187-194). A distinctive principle in the planning and construction of the settlements was the intended time period for which they were supposed to serve the residents. The time period was connected to the planned utilisation of forest resources in the area of a wood-harvesting unit. The instructions included a categorisation of different types of settlements.

They also defined types of housing, land use, services and infrastructure by settlement types.

In the settlement of Koivuselkä in particular, the enterprise built wooden barracks and houses almost at once in the early 1950s, to be used for between 15 and 20 years. Since this time, there has been little new residential construction by the forest enterprise. Furthermore, according to the residents, repairs to the houses and apartments have not necessarily met their needs and requirements.

Many of the original houses built by the logging enterprise are still in use today in both of the settlements. Amenities such as running water, sewerage and indoor bathrooms and toilets are not available in the settlements. The houses themselves are heated with wood. However, fetching and carrying both this fuel and household water creates problems, particularly for elderly residents. The need for major repairs to the residential accommodation and the construction of new houses is great.

Changes in the institutional providers for the settlements

The conditions of social infrastructure for forest sector are considered problematic in the current period of modernisation and reorganisation of production activities. In addition to the problems regarding enterprises, the work processes themselves and economic conditions, Strakhov *et al.* (1996, p.92-94) also refer to problems relating to the low levels of income and material well-being of the workers, the low level of services and medical care available to the population, the unsatisfactory housing conditions, the socio-spatial isolation of settlements and the poorly developed means of communication as having a negative impact on the social development taking place in small settlements engaged in the forest sector.

Many of the social responsibilities of the formerly state owned forest enterprises have already either been transferred to the local authorities, privatised or, in some cases, their provision has ceased altogether. However, the conditions regarding the provision of settlements can be nothing but clear in many places when the enterprises try to transfer the rest of the social responsibilities to the local authorities. Local authorities are unwilling to take on the responsibility of dilapidated infrastructure and housing especially when it is situated in small and remote settlements. The lack of financial and labour resources and thus the ability of the local authorities to maintain the provision are very limited.

The ties between the settlements of Koivuselkä and Matrosy and their former main provider, the forest enterprise, are disintegrating. Some of the services such as shops and canteens are now run by private trading companies or private entrepreneurs. Some, such as health care and kindergarten provision, have been transferred into the hands of the local authorities. Furthermore, owing to the decline in population and the reduced economic bases, many services, such as the company farm and some shops in Koivuselkä and the former company sauna in Matrosy, have been completely closed.

At the same time there have been major changes in the organisation of the activities of the department of service and housing provision of Shujales-company during the 1990s. More than 400 people were employed in the department in 1990. Today, there is only one foreman and one carpenter. However, the company is still responsible for the maintenance of housing in several settlements. (Pladov, 1998.)

Regarding the provision of residential housing, the conditions are somewhat different in the settlement of Koivuselkä and Matrosy (Tables 21.4 and 21.5). The bulk of the residents still live in company houses and apartments in Koivuselkä. The company has not succeeded in its negotiations with the local authorities of the Prjazha district concerning the transfer of the residential housing in this more remote settlement. The possibility of privatisation of apartments and garden plots has transferred more of the responsibility for accommodation to the residents themselves.

Table 21.4   Ownership of residence in Koivuselkä, (%)

Owner	%
Shujales-company	66
Private (self-built and privatised houses)	27
Owner not known	4
Local administration	3
Total number of interviewed households	112

Source: Household interviews in July 1994.

In Matrosy the privatisation of residences and, in particular, garden plots, have been more popular. The transfer of company houses to the local authorities in the Prjazha district took place in 1995. However, this transfer has not abolished the problem of poor physical living conditions in Matrosy. The resources available to the local authorities to maintain the provision, to repair and build houses with modern amenities are also very limited.

Table 21.5   Ownership of residence in Matrosy, (%)

Owner	%
Administration of Prjazha district	25
Private (self-built and privatised houses)	73
Other owner	2
Total number of interviewed households	105

Source: Household interviews in July 1997.

Reactions of residents

As a result of the decline in logging operations only a small number of people are now employed in forest work in the villages of Koivuselkä and Matrosy. The population has aged. Many of those who did not move out after the logging operations started to decline in the villages are former forest workers who are now pensioners. In addition to the diminished demand for labour in the forest work there are few employment opportunities available from the few local services. In this respect Matrosy has faired better. The employment opportunities have improved as a result of the relocation of a psychiatric hospital from Petrozavodsk in 1985, which employs 300 employees. However, part of this labour force is composed of employees who were with the hospital when it was located in Petrozavodsk and who now commute daily to and from Matrosy.

On the whole, the residents of these settlements are now more self-sufficient, relying on small-scale subsistence agriculture and the utilisation of berries and other natural products to provide their food supplies. Such activities are not new in Russia, but it is estimated that during the social and economic uncertainties brought by the transition processes they have become more important in the process of surviving everyday life (Varis, 1996, p.26-27). These activities reinforce the importance of the settlements themselves as the providers of livelihoods at a time when salaries and pensions are not enough and there are frequent delays in payment.

The notion of self-sufficiency has also made itself physically evident in the settlements. Since the beginning of the forestry settlements the company barracks and apartments were often regarded as uncomfortable and over-crowded for families. The standard housing provided by the enterprise was not always considered satisfactory in its ability to meet the needs and customs of the residents. Many have since built their own housing (see Tables 21.4 and 21.5). Even where the residents have not done this there have been several instances where they have constructed their own cattle sheds, huts, green houses and saunas near the housing supplied by the enterprise to complement the standard provision. The land, which is not used in this manner, is used intensively for growing food. In general, self-built private houses have been the usual form of tenure of the Soviet housing model in the rural areas (Andrusz, 1984, Kalinina, 1992).

Only a small number of the company apartments in Koivuselkä have been transferred from the enterprises to the local authorities so that the local schoolteachers may use them. However, the most notable difference between the villages is that in Matrosy the bulk of the residents live in private houses - either originally self-built and owned or privatised during the 1990s. In the village of Koivuselkä the bulk of residents still live in the company apartments. Apartments in semi-detached houses and barracks are not really attractive propositions for privatisation. Residents themselves can usually only afford superficial repairs and building work to be done. They usually carry this out themselves according to their abilities and resources. Therefore, few will wish to take on the burden that the privatisation of accommodation in such poor condition will entail.

The element of self-sufficiency found in the rural settlements is also important to those who are not residents of the settlements but who, nevertheless, have strong relationships with them through the existence of social networks. Close relatives, such as the children and grandchildren of the permanent residents, come to spend as much as several months with their older relatives in the villages during the summer periods to help with the garden plots and the collection of berries and mushrooms. They bring materials and labour with them from the town and are given locally produced food for their own households when they return to their own homes for the winter. Furthermore, during the winter months some pensioners live in the town only to return for the summer.

The future prospects for the settlements

Logging operations still continue to some extent in both of the settlements, even though the circumstances for the future of these operations are not promising. The options available regarding the future of the villages are tied up with whether their respective economic bases will continue to decline or whether the settlements will succeed in finding an alternative role. When this point was raised in the interviews with the residents they were seen to hold a different opinions on the subject (Table 21.6).

Table 21.6   Residents' opinions about the future of the home village, (%)

	Matrosy	Koivuselkä	
	1997	1994	1997
The village will 'die'	16	37	63
The village will survive and develop	4	12	1
The village will change into a dacha-village	66	9	4
No change	7	18	21
Can not say	7	25	11
Total	100	100	100
	n=105	n=112	n=72

Source: Household interviews in Koivuselkä and Matrosy

In Koivuselkä the residents see the future almost as having no prospects or potential for positive development. The residents' opinions in October 1997 compared with the opinions registered three years earlier provide a further illustration of just how bleak the situation of this remote settlement is felt to be. In 1994 an unemployment rate of 20% was calculated amongst the residents questioned during the households interviews (Klementev *et al.*, 1996, p.201).[2] Since then only one new shop representing new economic activity has appeared in the village.

In spite of the visiting summer residents in the households, a remote settlement such as this does not represent an ideal location for large-scale summer residency. However, a few families have moved from the town to Koivuselkä during the last few years in the belief that this rural settlement may provide a better livelihood through the subsistence activities taking place than the town even though there is little formal employment available in the settlement (Klementev *et al.*, 1996, p.202).

In Matrosy the situation regarding employment in forestry activities is also troubled. An unemployment rate of 26% was calculated amongst the members of the households which were interviewed in July 1997.[2] However, residents' opinions about the future prospects of the settlement are more positive as a place for summer residency. The settlement is located by the St. Petersburg - Murmansk highway and is quite near to Petrozavodsk, this makes it more easily accessible in comparison with

Koivuselkä. A considerable proportion of the residential housing is only in temporary use anymore. The former residents, who have already moved to the town, spend summer periods in their former home village. Three summer residence areas, dacha-co-operatives, add further to the number of non-permanent residents in the village.

Conclusions

The physical location of the two settlements seems to have a distinctive impact on their future prospects. The more remote settlement of Koivuselkä faces greater difficulties in finding an alternative role for the future to prevent the decline that is occurring in the economic base of the village. The settlement of Matrosy is in a better position to benefit from its physical location, even if this may mean that the village will have a substantially different role to that it has played in the past. This settlement is an attractive location for summer residences and it is well suited to meet the demands for self-sufficiency, which are present as a result of the uncertain socio-economic conditions brought by the current transition period in Russia.

Comparing the settlements in this manner illustrates the need to give Humbreys' (1988, p.97) proposition, that also remote resource communities with distinctive characteristics should be regarded as a part of the whole settlement system of an area rather than being considered in isolation.

The preliminary analysis of the two settlements also illustrates that in spite of their very similar roles during the era of Soviet central planning differences are also visible. The settlements are products of different time periods in the development of the forest sector in the area. In the settlement of Koivuselkä the influence of the earlier temporality in the settlement construction is more clearly visible as a starting point for the contemporary changes. The two settlements have different local starting points from which the changes in residence and settlement issues are developing.

The future task of this study will be to give further concentration to the human factor, the reactions and actions of residents and different resident groups regarding local community formation through social networks and identity of place around work and residence, through concrete everyday interaction between people and togetherness with the group in consciousness. In the Republic of Karelia one can ask: how will the post-socialist actions and choices differ from the past residential histories, and affect the community construction?

The final issue is to try to evaluate the potential the conditions of the Russian North offer for development. One can ask whether the transition will offer a starting point and the possibility of creating new, flexible and wanted practices by residents. This will lead us to consider questions of what possibilities do exist to implement a process of modernisation in relation to issues of residence and local community formation.

Notes

1 The source material consists of structured household interviews in the forestry settlements of Koivuselkä and Matrosy. In July 1994, 112 households were interviewed in Koivuselkä comprising 65% of the residents in the village. In September 1997, 72 households were interviewed. In July 1997, 105 households in Matrosy were interviewed comprising 25% of the residents in the village. In October 1997, 39 households were interviewed. Intensive theme interviews of key persons and representatives of organisations were also carried out. In addition to this the research also drew information from statistics, maps, pictures and observations.
2 Information is based on the classifications of the interviewees from the household interviews about the household members' situation regarding work.

References

Andrusz, G.D. (1984), *Housing and Urban Development in the USSR*. Studies in Soviet History and Society, MacMillan, London, p.354.

Autio, S. (1997), *Kohtalona metsä. Neuvosto-Karjalan metsätalouden murros v. 1926 - 1932*, A licenciate thesis of general history, University of Tampere.

Carter, J. and Jones, T. (1989), *Social Geography. An introduction to Contemporary Issues*, Edward Arnold, a division of Hodder and Stoughton, London, p.260.

Humbreys, J.S. (1988), 'Planning for Remote Settlements - People, Place and Prosperity', in T.B. Brealey, C.C. Neil and P.W. Newton (eds.), *Resource communities. Settlement and workforce issues*, CSIRO, Australia, pp.97-112.

Kalinina, N. (1992), 'Housing and Housing Policy in the USSR', in B. Turner, J. Hegedüs, and I. Tosics (eds.), *The Reform of Housing in Eastern Europe and the Soviet Union*, Routledge, London, pp.245-275.

Kemeny, J. (1992), *Housing and Social Theory*, Routledge, London, p.183.

Klementev, J. and Kozhanov, A. (1988), *Sel'skaja sreda i naselenie Karelii 1945-1960*, Nauka, Leningrad, p.211.

Klementev, J., Oksa, J. Polevshchikova, N. Rannikko, P. and Varis, E. (1996), 'Local impacts of restructuring. A case study of a forestry settlement', in E. Varis and S. Porter (eds.), *Karelia and St. Petersburg. From Lakeland Interior to European Metropolis*, University Press, Joensuu, Joensuu, pp.191-205.

Kortelainen, J. (1996), *Tehdasyhdyskunta talouden ja ympäristötietoisuuden murrosvaiheissa*, Publications in social sciences, N:o 99, University of Joensuu, Joensuu, p.155.

Myllynen, A-L., and Saastamoinen, O. (1995), 'Karjalan tasavallan metsätalous', *Silva Carelica* 29, Faculty of Forestry, University of Joensuu, Joensuu, p.210.

Oska, J. and Saastamoinen, O. (1985), 'Cross-Border Interaction and Emerging Interest Conflicts in the Forest Sector of Russian Karelia', in M. Dahlström, H. Eskelinen and U. Wiberg (eds.), *The East-West Interface in the European North*, Nordisk Samhällsgeografisk Tidskrift, Stockholm, pp.97-109.

Oska, J. and Varis, E. (1994), 'Karelian Republic: Population, Settlements and Administration', in H. Eskelinen, J. Oska and D. Austin (eds.), *Russian Karelia in search of a new role*, Karelian Institute, University of Joensuu, Joensuu, pp.57-69.

Pallot, J. (1987), 'Continuity and change in village planning from the 18th century', in L. Holzner, and J.M. Knapp (eds.), *Soviet Geography Studies In Our Time*, The College of Letters and Science, The American Geographical Society, Collection of the Golda Meir Library, The University of Wisconsin-Milwaukee, Milwaukee, pp.319-349.

Pladov, V. (1998), 'Director, AOZ Shujales, personal communication 15.6.1998', Chalna.

Rannikko, P. (1989), *Metsätyö-pienviljelykylä. Tutkimus erään yhdyskuntatyypin noususta ja tuhosta*, Publications of social science, N:o 12, University of Joensuu, p.114.

Respublika Karelija v tsifrah (1997), Gosydarstvennyj komitet Respubliki Karelija postatistike, Petrozavodsk.

Souza de, P. (1989), Territorial Production Complexes in the Soviet Union - with a special focus on Siberia, *Departments of Geography, Serie B nr* 80, University of Gothenburg, p. 257.

Strakhov, V., Teplyakov, V.K. Borisof, V.A. Goltsova, N.I. Saramäki, J. Niemelä P. and Myllynen, A-L. (1996), *On the Ecological and Economic Impacts of Wood Harvesting and Trade in North-West Russia*, OY FEG - Forest and Environment Group Ltd., Joensuu, p.152.

Tehnicheskie ykazanija. Po proektirovaniju lesozagotovitel'nyh predprijatij (1964), Gosydarstvennyj komitet po lesnoj, tseljulozno-bumaznoj, derevoobrabatyvajushchej promyshlennosti i lesnomy hozjajstvy pri Gosplane SSSR, Leningrad.

Tykkyläinen, M. (1994), *Kaupunkilaismainarit Forrestaniassa. Työpaikkamajoituksen vaikutus aluerakenteeseen Länsi-Australiassa*, University Press, Joensuu, p.214.

Tykkyläinen, M. (1998), 'Theoretical and methodological underpinnings of the study of rural survival strategies in transitional countries,' in M. Tykkyläinen, E. Varis, J. Oksa, M. Piipponen, I. Nagy, É. Kiss and G. Mátray, *Rural survival strategies in transitional countries. An introduction to the comparative study of localities in Northwestern Russia and Hungary*, University of Joensuu, Karelian Institute, Working papers N:o 2/1998, Joensuu, pp.5-14.

Tykkyläinen, M. (1999), 'The emergence of capitalism and struggling against marginalization in the Russian North', in H. Jussila, R. Majoral and C. Mutambirwa (eds.), *Marginality in Space – Past, Present and Future*, Ashgate, Aldershot.

Tykkyläinen, M. and Neil, C. (1995), 'Socio-Economic Restructuring in Resource Communities: Evolving a Comparative Approach', *Community Development Journal*, vol. 30, pp.31-47.

Varis, E. (1996), *The Restructuring of Peripheral Villages in Northwestern Russia, Research for Action*, The United Nations University, WIDER, Helsinki, p.37.

Part 4 – Summary and conclusions

22 Summary and conclusions

ROSER MAJORAL AND HEIKKI JUSSILA

This book of *Environment and Marginality in Geographical Space* has analysed the issue from various points of view. The book starts with a theoretical article about the importance of land use for man. The fact that land as such has been and is one of the fundamentals of being in this Earth of ours makes the first article of this book very interesting indeed. The book then goes on to look at the issues of environment in the globalizing world through the 'spectaculars' of economic and social geography where the interplay between man and nature is always present, although the role nature in this context has been changing dramatically during the last decades. Today the questions of global warming and the changes that man has provoked in atmosphere, the green house gases, are coming more apparent and these phenomena have profound effects on the life in rural areas that constitute the most 'green areas' of the world. The economic problems arising from environmental issues coupled with those of the globalization are the issues that this book looks into.

The first part of the book discusses the issues of *Land Use and Environment*. In this section of the book the first chapter 'Land use and abuse' by Leimgruber discusses about the importance of land use and how man has been conscious of the importance of land, nature and environment to his activity and survival through history and how land use decisions were and are still guided by the prevailing worldview of a society that was originally directed toward a harmonious, sustainable coexistence with the economic system, but which is now dominated by technical considerations. He calls for a new world view that leads us back to the moral order, to a consciousness of what is right for humanity. A shift back on the value continuum from secular toward sacred values seems necessary.

In his paper on trajectories of change in marginal and critical regions of Southeast Asia and Southern China, Hill hypothesises a number of trajectories of change using a time frame of about 30 years: a continuance of the status quo in which people and resources are more or less in balance; a downward spiral initiated by soil erosion on intensively cultivates sloping lands, decreasing forest or scrub fallows, over-exploitation of forest in the face of continued population growth and few employment opportunities outside the local economy; and finally a possible upward spiral that would include tourism and other activities controlled by marginal peoples themselves although this outcome is likely to require affirmative governmental action and excess capital, skills and other resources not likely to be available to most of the foreseeable future.

The tremendous growth of population and its increasing pressure on agricultural land is the departing point for Singh's paper 'Economic scenario of the reclamation and utilisation of marginal lands in India'. Reclamation of marginal lands is seen by the author as one of the ways for improving the socio-economic conditions of poor by providing more cultivated and pasture land, more employment opportunities and increase production and incomes of rural people. This would lead to an increase of the size of holdings.

Four chapters are devoted to different Argentinean situations: Gutiérrez and Furlani analyse in their paper on Agro-industrial enterprises and environmental fragility, the society and territories transformation due to processes connected with economic opening, which have accelerated technical advances, internal decentralisation policies and the consolidation of country blocs. The area of Mendoza is studied as the tendencies of foreign trade and the concerns, motivations and attitudes of agro-industrial enterprises, necessary agents of an activity change. The lack of complementation between external dynamics and territorial values leads to endogenous development as an attitude that favours local growth of culture and environment is as well examined. The study has been carried in a reduced but significant area, marginal by its ecological risks.

Always in the area of Mendoza (Uco Valley in this occasion), Molina examines in her paper on the problem of decision making within a fragile ecosystem, how the territorial fragmentation, due to multiple influences, diversify and produce overlapping decisions of different institutions, working independently one from the other, on services supply, natural resources, production and social needs. All this results in a complex environmental planning and overload of the duties and bureaucratic burdens.

Schmite contribution on land use in the Argentinean Pampa explains how the current world situation requires that La Pampa agricultural producers seek for commercial insertion ways that are increasingly competitive producing alterations in the productive activities and the traditional social structure. New actors with a different productive logic intervene in the agrarian system, generating changes with a high degree of dynamism.

Complementing the previous contribution Tourn deals with the influence of globalization to the Argentinean Pampa. The growing external influence has brought in new interests to the region. These interests have produced a technological change on the information flows and there is currently an ongoing increase of direct private investments on productive systems. According to Tourn these elements have had negative effects on the environment through land-use by increasing land deterioration as well as on the regional economy by enlarging territorial disparities within the region.

The chapter of Geist on 'Transforming the fringe: Tobacco-related wood usage and its environmental implications', uses a crop specific approach by addressing the notion of felling trees in the special case of tobacco as the farming of the crop seems to become increasingly understood as one of the major social and economic deriving forces of global environmental change since about the start of commercial growing. The significance of tobacco farming as it mainly occurs at the agriculture/forest frontier in growing countries of the developing world within the present context of efforts aimed at regulating the global tobacco industry. Major emphasis is put upon empirically derived dimensions of recent wood usage for the curing of the crop and the resulting consequences upon land use and land cover changes derived.

The issue of the *environment* in this book does not only concentrate on the issues of 'pure' environment. The topic has been understood with a wider perspective and consequently this volume contains articles that do not meet the 'strict' definition of 'environmental issues'. However, it is important to understand that environment is also an economic and social phenomenon that influences to the use of natural resources and through this also indirectly to the possibilities to avoid environmentally harmful practices.

The question of *environment* is present in all the articles of the second section. The important message from all of them is that to humans 'environment' is much more than just nature, although without clean nature life is not possible. The role of various obstacles, e.g., borders as presented by Capella and Font or Tort, or the issues of social barriers produced by economic or health conditions (Blom and Nossa), all point out that the question of how we are utilising our resources is important and in this respect it is not only important to overcome problems of pollution but the issues of social nature need to be resolved also.

It is in this light that *Territorial Marginalization* is looked at in the second part of the book from different points of view and scales. It starts with

Fernandes' paper about the concept of territorial marginality, taking the Portugal image at the end of the 20th century as a standpoint. Fernandes emphasises how difficult is to concretise the term and the criteria to be used to define the concept. His observations can be demonstrated in the two contributions that follow, dealing with territorial marginality in Spain, each using different criteria for defining the concept, within on a large region, Galicia and the second on a small territories in Catalonia.

Galicia is seen by Piñeira and Rodríguez as a peripheral and disadvantaged region in Spain although a process of modernisation and urbanisation is going on both in society, productive and territorial structures. Stance and development structures are actually organised in terms of its integration in the EU, which financial support is fundamental to counter counteract Galicia's peripheral structure. On the other hand Cors, in his paper on his study case on reality against perception of marginality, states the relativeness of the concept and how it deepens on the scale of analysis which is applied. In this way he insists in how some rural areas perceived as traditionally marginal have no longer this nature. His case study demonstrates how the growing interest in discovering the countryside by urban people has lead to the gradual disappearance of the pejorative connotations traditionally associated with marginal areas.

Two contributions deal with borders as marginal territories in Spain. The article of Capella and Font about regional borders and globalization, explain how frontiers frequently constitute marginal zones due to physical or political reasons and how new political and economic situations may led to suppress or strength the borders and so the marginality of the surrounding territories. They take as an example the internal and external frontiers of the EU and the regional borders within Spain.

A more detail discussion of this is given in the article by Tort about the border between the autonomous communities of Aragón and Catalonia, fruit of a political compromise, that has lead to the emergence of a sense of marginality among the people due to the anomalous position of the border which runs counter to their everyday needs and interests.

The question of territorial marginalization moves from a larger geographical region into a city-region in the chapter by Blom on 'Microphery in Megacities'. He stresses the idea that frequently wealthy inner city centres and poor suburban ghettos exist irrespective either in small urban areas or megacities becoming increasingly apparent that we have got a society where no matter whether they are defined on economic social, ethnic or demographic principles, different groups of people tend to gather in different housing states without contact with one another. The author has taken twelve communities in Manhattan (NY) as an example for discussing the differences and similarities between adjoining urban districts from a centre periphery perspective.

The second part ends with a contribution on the world distribution of AIDS/HIV, written by Nossa. He states that although third world countries and the poorer and less educated people are feeling the brunt of the epidemic, it has spread in different ways and through different groups of people in different parts of the world. From a territorial point of view, neighbouring countries often have very different epidemics and even within a single country they can strike different populations or different geographical areas in dissimilar ways, ways that may change over the course of time. He concludes that educational programs are fundamental as educated people have better access to information and how only a collective effort can help to delete the big gap of inequality between people and countries.

The last part of this book looks at *Development in the Margins and Peripheries*. This part takes the question of environment and looks at it from economic and regional points of view. The issue is that of the 'whole', i.e., one analyses the system and not just one aspect of it. It is for this reason that the contributions in the last part of the book analyse, in general terms, peripheries and margins and their development possibilities.

The two first chapters are about changes and alternatives in country peripheries. Maas and Wisserhof work, focuses its attention on aspects of scale of pink (pig) invasion in the Dutch periphery, particularly the rescaling of the social activities involved. Besides an increase of scale (internationalisation and globalization) a decrease of scale (regionalisation and localisation) also occurs in Dutch agriculture. Arable farming in peripheral regions is increasingly grounded on regional and local impulses and resources in part due to the pig invasion to those regions. Despite the relatively marginal economies of some regions the strongly resist to de pink invasion and a number of small-scale regional alternatives to large-scale pig farming are being developed. Even peripheral regions are able to develop endogenous alternatives to exogenous impulses.

Grossman and Moshayov, on their turn, look at the agrarian structure of Israel's southern periphery. Studying the Western Negev, they demonstrate that technology is capable of altering the status of marginal zones and may even have the potential of turning the periphery into an economic core. The contraction of the area available for agriculture in the country's core zone has shifted parts of the farming activity to the Israeli periphery where is little competition from urban-industrial users. The new trends in this peripheral area, associated with the demise of agriculture in the core, may result in a second revolution, which will drastically alter the landscape, the pattern of agricultural land use and the population.

While the first chapters in this section speak about farming and agriculture. The article of Tykkyläinen that looks at the issues of northern Russia analyses the issues of industrial development. The economic changes that have taken place in there have radically changed the possibilities of local actors to cope with both national and global challenges. The approach of the article is micro-scale, an analysis of two firms and their possibilities to cope with the economic crisis. The article shows clearly that currently the Russian firms do not have possibilities to do real development and economic planning since all the incomes they have go almost directly to covering wages as well as (old) social obligations, leaving very little for long-term investments.

The social atmosphere in the region is also not favourable for new type of entrepreneurial activities. People do not have the culture to act according to business ideas. The interplay between human agency and economic structures that leads to development is weakly present in north-western Russia. This means that economic development in Russia is going through a unique form of transition, currently consisting a combination of laissez-faire, conventional market economy as well as modernised socialistic practices. In this kind of environment the globalization of economy leads to pockets of development that are very small and as a whole does not influence the economic development within the region. The old socialist structures of Russia also keep the pace of development low and also for this reason the globalization of economy by the arrival of external investors is probably the only solution in order to keep the current firms of the region functioning.

Settlement issues are the concern of the two following papers by Varis and Piipponen, dealing both with Russian Karelia. They analyse at different scales the restructuring of rural settlements in the transition of post-socialist countryside. Continuous changes have created a situation in which the local people must create their survival strategies to adapt to the changes. In spite of their similar roles during the era of Soviet central planning differences are visible. The authors wonder if the transition will offer a starting point and the possibility of creating new flexible and wanted practices by residents.

The articles in the last section of this book have looked at the question of 'environment and land use' in marginal areas and the point of view has been most of all economic and social. The issue that arises from the discussion in this last section is similar to the one on *Territorial Marginalization*.

The needs of people are different and consequently the use of land or 'the environment' depends on those needs. The articles regarding farming look at alternatives for marginal regions or 'environments'. The article of Tykkyläinen analyses the issue of industry in a marginalized situation of Russia. The issues raised in the last two articles by Varis and Piipponen come again closer to the social issues discussed, e.g., by Blom and Nossa, although they use settlement approach.

The articles in this book seek to understand the various 'forces' that shape the geographical space where people work and life. It is this that links them to the other articles; the use of resources is and has been both the reason and the cause for environmental problems. The questions raised and discussed in this volume point all to the same direction, it is important that the 'whole' of geographical space is been taken into account when looking at the issues of the 'environment'. Land has been and continues to be important to people but as the standard of living has risen so have changed the needs of people. The lesson is that besides the issues of nature one should and must also take care of the issues of people, since only in this way it is possible to face the difficult problems of environment that face us on this new Millennium.